An Assessment of NASA's National Aviation Operations Monitoring Service

Committee on NASA's
National Aviation Operations Monitoring Service (NAOMS) Project:
An Independent Assessment

Aeronautics and Space Engineering Board

Division on Engineering and Physical Sciences

NATIONAL RESEARCH COUNCIL
OF THE NATIONAL ACADEMIES

THE NATIONAL ACADEMIES PRESS
Washington, D.C.
www.nap.edu

THE NATIONAL ACADEMIES PRESS **500 Fifth Street, N.W.** **Washington, DC 20001**

NOTICE: The project that is the subject of this report was approved by the Governing Board of the National Research Council, whose members are drawn from the councils of the National Academy of Sciences, the National Academy of Engineering, and the Institute of Medicine. The members of the committee responsible for the report were chosen for their special competences and with regard for appropriate balance.

This study is based on work supported by Contract NNH05CC16C between the National Academy of Sciences and the National Aeronautics and Space Administration. Any opinions, findings, conclusions, or recommendations expressed in this publication are those of the author and do not necessarily reflect the view of the agency that provided support for the project.

International Standard Book Number-13: 978-0-309-14646-3
International Standard Book Number-10: 0-309-14646-1

Cover: Image courtesy of My Brainstorm Center, available at www.mybrainstormcenter.com. Design by Tim Warchocki.

Copies of this report are available free of charge from:

Aeronautics and Space Engineering Board
National Research Council
500 Fifth Street, N.W.
Washington, DC 20001

Additional copies of this report are available from the National Academies Press, 500 Fifth Street, N.W., Lockbox 285, Washington, DC 20055; (800) 624-6242 or (202) 334-3313 (in the Washington metropolitan area); Internet, http://www.nap.edu.

Printed in the United States of America

OTHER REPORTS OF THE AERONAUTICS AND SPACE ENGINEERING BOARD AND THE SPACE STUDIES BOARD

America's Future in Space: Aligning the Civil Space Program with National Needs (Space Studies Board [SSB] with the Aeronautics and Space Engineering Board [ASEB], 2009)
Approaches to Future Space Cooperation and Competition in a Globalizing World: Summary of a Workshop (SSB with ASEB, 2009)
Assessment of Planetary Protection Requirements for Mars Sample Return Missions (SSB, 2009)
Fostering Visions for the Future: A Review of the NASA Institute for Advanced Concepts (ASEB, 2009)
Near-Earth Object Surveys and Hazard Mitigation Strategies: Interim Report (SSB with ASEB, 2009)
A Performance Assessment of NASA's Heliophysics Program (SSB, 2009)
Radioisotope Power Systems: An Imperative for Maintaining U.S. Leadership in Space Exploration (SSB with ASEB, 2009)

Assessing the Research and Development Plan for the Next Generation Air Transportation System: Summary of a Workshop (ASEB, 2008)
A Constrained Space Exploration Technology Program: A Review of NASA's Exploration Technology Development Program (ASEB, 2008)
Ensuring the Climate Record from the NPOESS and GOES-R Spacecraft: Elements of a Strategy to Recover Measurement Capabilities Lost in Program Restructuring (SSB, 2008)
Final Report of the Committee for the Review of Proposals to the 2008 Engineering Research and Commercialization Program of the Ohio Third Frontier Program (ASEB, 2008)
Final Report of the Committee to Review Proposals to the 2008 Ohio Research Scholars Program of the State of Ohio (ASEB, 2008)
Launching Science: Science Opportunities Provided by NASA's Constellation System (SSB with ASEB, 2008)
Managing Space Radiation Risk in the New Era of Space Exploration (ASEB, 2008)
NASA Aeronautics Research: An Assessment (ASEB, 2008)
Opening New Frontiers in Space: Choices for the Next New Frontiers Announcement of Opportunity (SSB, 2008)
Review of NASA's Exploration Technology Development Program: An Interim Report (ASEB, 2008)
Science Opportunities Enabled by NASA's Constellation System: Interim Report (SSB with ASEB, 2008)
Severe Space Weather Events—Understanding Societal and Economic Impacts: A Workshop Report (SSB, 2008)
Space Science and the International Traffic in Arms Regulations: Summary of a Workshop (SSB, 2008)
United States Civil Space Policy: Summary of a Workshop (SSB with ASEB, 2008)
Wake Turbulence: An Obstacle to Increased Air Traffic Capacity (ASEB, 2008)

Limited copies of SSB reports are available free of charge from

Space Studies Board
National Research Council
The Keck Center of the National Academies
500 Fifth Street, N.W., Washington, DC 20001
(202) 334-3477/ssb@nas.edu
www.nationalacademies.org/ssb/ssb.html

COMMITTEE ON NASA'S NATIONAL AVIATION OPERATIONS MONITORING SERVICE (NAOMS) PROJECT: AN INDEPENDENT ASSESSMENT

VIJAYAN N. NAIR, University of Michigan, Ann Arbor, *Co-chair*
CLINTON V. OSTER, JR., Indiana University, Bloomington, *Co-chair*
DAVID L. BANKS, Duke University
ROBERT M. BELL, AT&T Laboratories
JOHNNY BLAIR, Abt Associates
ANTHONY J. BRODERICK, Independent Consultant
JAMES DANAHER, National Transportation Safety Board (retired)
PETER GRIFFITHS, International Air Transport Association
IAIN M. JOHNSTONE, Stanford University
KAREN KAFADAR, Indiana University, Bloomington
ELIZABETH A. LYALL, Research Integrations, Inc.
DONALD W. RICHARDSON, Donrich Research, Inc.
THOMAS B. SHERIDAN, Massachusetts Institute of Technology (emeritus)
ALFRED T. SPAIN, JetBlue Airways Corporation (retired)
S. LYNNE STOKES, Southern Methodist University

Staff

PAUL JACKSON, Study Director, Aeronautics and Space Engineering Board
NEAL GLASSMAN, Senior Program Officer, Board on Mathematical Sciences and Their Applications
ANDREA REBHOLZ, Program Associate, Aeronautics and Space Engineering Board (from February 2009)
SARAH CAPOTE, Program Associate, Aeronautics and Space Engineering Board (through November 2009)
LEWIS GROSWALD, Research Associate, Space Studies Board

AERONAUTICS AND SPACE ENGINEERING BOARD

RAYMOND S. COLLADAY, Lockheed Martin Astronautics (retired), *Chair*
KYLE T. ALFRIEND, Texas A&M University
AMY L. BUHRIG, Boeing
PIERRE CHAO, Center for Strategic and International Studies
INDERJIT CHOPRA, University of Maryland
JOHN-PAUL B. CLARKE, Georgia Institute of Technology
RAVI B. DEO, Northrop Grumman Corporation (retired)
MICA R. ENDSLEY, SA Technologies
DAVID GOLDSTON, Harvard University
JOHN HANSMAN, Massachusetts Institute of Technology
JOHN B. HAYHURST, Boeing Company (retired)
PRESTON HENNE, Gulfstream Aerospace Corporation
RICHARD KOHRS, Independent Consultant
IVETT LEYVA, Air Force Research Laboratory
ELAINE S. ORAN, Naval Research Laboratory
ELI RESHOTKO, Case Western Reserve University
EDMOND SOLIDAY, United Air Lines (retired)

Staff

RICHARD E. ROWBERG, Interim Director (from March 2, 2009)
MARCIA S. SMITH, Director (through March 1, 2009)

COMMITTEE ON APPLIED AND THEORETICAL STATISTICS

KAREN KAFADAR, Indiana University, *Chair*
AMY BRAVERMAN, Jet Propulsion Laboratory
CONSTANTINE GATSONIS, Brown University
MICHAEL F. GOODCHILD, University of California, Santa Barbara
MICHAEL NEWTON, University of Wisconsin-Madison
MICHAEL STEIN, University of Chicago

Staff

SCOTT T. WEIDMAN, Director
NEAL GLASSMAN, Senior Program Officer

Preface

On October 22, 2007, a news article from the Associated Press (AP) brought a relatively obscure NASA project, the National Aviation Operations Monitoring Service (NAOMS)—a survey administered to pilots from April 2001 through December 2004—to the attention of the public as well as Congress.[1] The article revealed that a letter from NASA had indicated that a Freedom of Information Act (FOIA) request by the AP for information related to the NAOMS project was being turned down. Specifically, the letter stated: "Release of the requested data, which are sensitive and safety-related, could materially affect the public confidence in, and the commercial welfare of, the air carriers and general aviation companies whose pilots participated in the survey."[2]

Citing an unnamed source familiar with the results of the survey, the AP article also reported that the survey data showed that "the pilots reported at least twice as many bird strikes, near mid-air collisions, and runway incursions as other government monitoring systems show."[3]

A few days before the AP news story broke, the Subcommittee on Investigations and Oversight of the U.S. House of Representatives' Committee on Science and Technology sent a letter to NASA Administrator Michael Griffin requesting that a variety of materials pertaining to the NAOMS project be turned over "to help the subcommittee understand more clearly what information NASA collected in the three years that it surveyed pilots in the NAOMS project."[4]

The Committee on Science and Technology held a hearing on NAOMS on October 31, 2007, at which Dr. Griffin expressed disagreement with the phrasing of the FOIA denial letter and firmly stated that NASA was not putting "commercial interests ahead of public safety."[5] According to Dr. Griffin, the FOIA request was denied because "the data likely contained confidential commercial information."[6] He further indicated that NASA would be releasing all NAOMS data that did not contain either confidential commercial information or information that

---

[1] Rita Beamish, "NASA Sits on Air Safety Survey," Associated Press, October 22, 2007.

[2] Thomas S. Luedtke, Associate Administrator for Institutions and Management, NASA, Letter to Adam J. Rappaport, Levine, Sullivan, Koch, and Schulz, L.L.P., September 5, 2007.

[3] Beamish, "NASA Sits on Air Safety Survey," 2007.

[4] Brad Miller, U.S. Representative, Chairman, Subcommittee on Investigations and Oversight, Committee on Science and Technology, U.S. House of Representatives, Letter to Michael Griffin, Administrator, NASA, October 19, 2007.

[5] Michael Griffin, Administrator, NASA, Statement before the Subcommittee on Investigations and Oversight, Committee on Science and Technology, U.S. House of Representatives, October 31, 2007, p. 4.

[6] Ibid., p. 3.

would compromise pilot identity. Redacted versions of the data were subsequently released to the public, with the first release occurring in December 2007.

Dr. Griffin also pointed out:

> None of the research conducted in the NAOMS project, including the survey methodology, has been peer-reviewed to date. Accordingly, any product of the NAOMS project, including the survey methodology, the data, and any analysis of that data, should not be viewed or considered at this stage as having been validated.[7]

The above statement speaks to the central purpose of this assessment. In fact, just prior to the congressional committee hearing on October 31, 2007, NASA had asked the National Research Council (NRC) of the National Academies to review the NAOMS survey methodology.

To conduct the review requested by NASA, the NRC established the Committee on NASA's National Aviation Operations Monitoring Service (NAOMS) Project: An Independent Assessment, composed of experts in the fields of aviation operations (including pilots), aviation safety, survey methodology, and statistics. Biographical information on the committee members is provided in Appendix A of this report.

The committee was asked to assess the NAOMS survey methodology and to analyze the publicly available survey data to determine their potential utility. Additionally, NASA requested that the committee provide recommendations on the most effective ways to use the NAOMS data. The committee's full statement of task is as follows:

> The Aeronautics and Space Engineering Board (ASEB), in conjunction with the Committee on Applied and Theoretical Statistics (CATS), will create an ad hoc study committee to make an independent assessment of NASA's National Aviation Operations Monitoring Service (NAOMS) project. The NAOMS project used a survey methodology to anonymously collect data from commercial and general aviation pilots over several years regarding aviation safety-related events. The NAOMS project contracted with Battelle Memorial Institute to design the survey and collect the data. The study committee will assess the NAOMS survey methodology, and, to the extent possible, analyze the survey data. This assessment will be based upon information in the public domain including the following items that will be provided to the committee by NASA: (1) a final report provided by the prime contractor, Battelle, that was released to the public on December 31, 2007; (2) a November 13, 1998, NAOMS briefing to the NASA Aviation Safety Reporting System Advisory Subcommittee; and (3) the redacted set of survey responses that were released to the public on December 31, 2007. The study committee will also provide recommendations on the most effective ways to use the NAOMS data.
>
> Specifically, as part of the assessment, the study committee shall:
>
> 1. Assess the process used by the contractor and described in the contractor report to determine how to acquire a statistically meaningful data set representative of a variety of factors that may affect (or impact) the safety of the national airspace system and that would enable one to track how these factors change over time.
> 2. Assess the advantages and disadvantages of using a survey method to collect such a statistically meaningful data set.
> 3. Assess the survey methodology used by the contractor and described in the contractor report to include:
>
> (a) An analysis of specific details of the survey methodology such as the recall period, collection approach, sampling approach, questionnaire design and the use of non-aviation experts as interviewers.
> (b) An analysis of method or methods used to validate the survey methodology.
> (c) An identification of the various sources of error (both random and systematic) due to the survey methodology, along with estimates of the magnitudes of those errors, including an analysis of the adequacy of the sample size.
> (d) Recommendations of how one might estimate appropriate error bars for the survey results.
> (e) Recommendations regarding any methods that might enable one to correct for errors introduced by the methodology.
>
> 4. Conduct an analysis of the project survey data provided by NASA to determine its potential utility. (Note: The survey data will be a redacted data set also released to the public. This data set will be redacted in a manner that

[7] Ibid.

preserves the anonymity of the pilot respondents. Details of the redaction process will be provided.) This analysis may include an assessment of the data's validity using other known sources of information.

5. Provide recommendations on the most effective ways to use the NAOMS data. Such recommendations can include the possibility of using the data in combination with other safety data.

Between June 2008 and March 2009, the committee held five meetings to receive briefings, review materials, and write this report. This schedule included a meeting at the NASA Ames Research Center, where a subcommittee met with some of the NASA researchers connected with the NAOMS project. During the course of the study, the committee received briefings from representatives of a variety of organizations on issues related to the NAOMS project. Appendix B lists these presenters.

This report presents the committee's analyses and findings. Following the Summary, Chapter 1 provides background on NAOMS and a general description of the features of the NAOMS survey. Chapter 2 discusses how aviation safety is measured, what other sources of aviation safety data existed when NAOMS was developed, and what data exist today. Chapter 3 describes how sample surveys are conducted, examines some of the major surveys used in the government sector, and assesses the advantages and disadvantages of using a survey in the field of aviation safety. Chapter 4 assesses the NAOMS sampling design; it examines the impact of the specific features of the sample design and the impact of potential coverage biases on the accuracy of the estimates that are based on these data. Chapter 5 examines the structure of the questionnaire used in the project as well as the content and wording of the specific questions. Chapter 6 explores the limitations imposed by the redacted data set. Chapter 7 summarizes the committee's analysis of the publicly available data set and provides its recommendation on the most effective ways to use the NAOMS data. In addition to Appendixes A and B mentioned above, Appendix C lists select acronyms used in the report, Appendix D discusses the principal segments of aviation in the United States, Appendix E gives additional examples of surveys by federal agencies, Appendix F lists the additional survey questions with problems discussed in Chapter 5, Appendix G reprints the full air carrier questionnaire, and Appendix H reprints the full general aviation questionnaire.

Acknowledgment of Reviewers

This report has been reviewed in draft form by individuals chosen for their diverse perspectives and technical expertise, in accordance with procedures approved by the Report Review Committee of the National Research Council (NRC). The purpose of this independent review is to provide candid and critical comments that will assist the institution in making its published report as sound as possible and to ensure that the report meets institutional standards for objectivity, evidence, and responsiveness to the study charge. The review comments and draft manuscript remain confidential to protect the integrity of the deliberative process. We wish to thank the following individuals for their review of this report:

Peter J. Bickel, University of California, Berkeley,
R. John Hansman, Massachusetts Institute of Technology,
Raymond LaFrey, Lincoln Laboratory (retired),
Daniel R. Masys, Vanderbilt University School of Medicine,
Edmond Soliday, United Airlines (retired),
Roger Tourangeau, University of Maryland, and
Kirk Wolter, National Opinion Research Center.

Although the reviewers listed above have provided many constructive comments and suggestions, they were not asked to endorse the conclusions or recommendations, nor did they see the final draft of the report before its release. The review of this report was overseen by Lawrence D. Brown, University of Pennsylvania, and Adib Kanafani, University of California, Berkeley. Appointed by the NRC, they were responsible for making certain that an independent examination of this report was carried out in accordance with institutional procedures and that all review comments were carefully considered. Responsibility for the final content of this report rests entirely with the authoring committee and the institution.

Contents

APPENDIXES

Summary

The National Research Council (NRC) of the National Academies was asked by the National Aeronautics and Space Administration (NASA) to perform an independent assessment of NASA's National Aviation Operations Monitoring Service (NAOMS) project, which was a survey administered to pilots from April 2001 through December 2004. To conduct this review, the NRC established the Committee on NASA's National Aviation Operations Monitoring Service (NAOMS) Project: An Independent Assessment, consisting of experts from the fields of aviation safety, aviation operations (including several pilots), survey methodology, and statistics. The committee reviewed various aspects of the NAOMS project, including the survey methodology, and conducted a limited analysis of the publicly available survey data.

Sample surveys have been used routinely by federal agencies to collect and analyze data in order to inform policy decisions and assess national needs. They can also be used effectively to provide statistically valid information on rates and trends of events (such as bird strikes or rejected takeoffs) that are potentially related to aviation safety. In this context, surveys have several advantages over other sources of data: for example, they could provide reliable information about all segments of civilian aviation and characterize the safety of general aviation (GA) flights and the safety of the flights of other segments of aviation where data are limited. Further, government-sponsored surveys can produce data that are accessible to the public and can be analyzed regularly and independently. However, past experience in the government sector indicates that successful large-scale surveys typically require a substantial commitment of time and resources to develop, refine, and improve the survey methodology and to ensure that the survey provides useful and high-quality data.

Several aspects of the NAOMS survey design were consistent with generally accepted practices and principles in survey design, and the committee finds these aspects to be reasonable and appropriate. These include the choice of a cross-sectional design, the computer-assisted telephone interview (CATI) method, and the use of professionally trained interviewers. A CATI system has the potential to incorporate checks for unlikely or implausible values during the interview process. However, the committee found that substantial fractions of the reported non-zero counts of events and reported flight legs and hours flown had implausibly large values, suggesting that the NAOMS survey did not take full advantage of this feature of CATI. The NAOMS team also faced challenges in the choice of the sampling frame and had to make compromises at several stages. Unfortunately, the use of the publicly available Airmen Certification Database for the sampling frame and the criteria used to draw the sample of pilots in the air carrier (AC) survey led to biases in the sample, with an over-representation of wide-body aircraft and an under-representation of small aircraft. While the choices and compromises by NAOMS may have been for good reasons, the team should have investigated their potential impact as well as the magnitude of biases resulting from failure to locate sampled pilots and other forms of nonresponse. In particular, the collection and analysis of supplemental data during the early phase of the survey would have enabled a reliable assessment of the various biases and may have led, if necessary, to the development of alternative design strategies.

The committee identified deficiencies in the structure and wording of the questions used in the survey. Some of the questions asked pilots for information that they would not have had without a post-flight analysis. Other questions had complex structure or multiple parts or used vague phrases to describe the events that the survey was attempting to measure. In addition, both the AC and the GA questionnaires asked respondents to include events, flight hours, and flight legs in segments of aviation that went beyond even the broadest definition of AC operations and beyond the conventional definition of GA operations. As a result, highly disparate segments of the aviation industry were aggregated into the safety-related event rates that were calculated from the AC and GA surveys. Finally, the inability to link safety-related events to the aircraft type or to the type of operating environment in which the event occurred severely hinders any meaningful analysis of event rates or trends in event rates by aircraft type or by segment of aviation.

The committee did not have access to the original survey data. The redacted data sets have several limitations that further constrain the ability to analyze the data to meet the committee's objectives. For example, the time of survey response was grouped into years, so estimates of event rates could be computed only by years. This limits the ability to track changes in event rates over shorter timescales, to determine the effects of changes in the aviation system on event rates, and to assess seasonal and similar types of effects. In addition, grouping the exposure data (number of hours and flight legs flown) into categories increases the uncertainty in the estimates of event rates broken down by key characteristics, such as pilot experience and plane type. Issues associated with preserving respondents' anonymity and confidentiality and with the public release of data have been known in the survey community for some time, and these issues should have been anticipated and addressed at the design stage of the NAOMS project.

The committee's limited analysis of the redacted data revealed serious problems with data quality: substantial fractions of the reported non-zero counts of events had implausibly large values, and respondents often rounded their answers to convenient numbers. The extent and magnitude of these problems raise serious concerns about the accuracy and reliability of the data. In the committee's view, some of these problems could have been reduced substantially if more effort had been spent on ensuring data accuracy during the interview and data-entry stages and if respondents had been asked to refer to their logbooks when possible. This would have been especially useful in providing reliable information on the number of hours flown and the number of flights (takeoffs/landings) and in helping to confine the answers to the recall period. The committee does note that many of the biases that are relevant for estimating event rates would be mitigated for trend analysis to the extent that the biases remain relatively constant over time. However, the degree of mitigation might vary substantially across event types.

The committee did not find any evidence that the NAOMS team had developed or documented data analysis plans or conducted preliminary analyses as initial data became available in order to identify early problems and refine the survey methodology. These activities should be part of a well-designed survey, especially a research study to assess the feasibility of survey methodology in aviation safety.

Given the deficiencies identified, and despite some methodological strengths of the NAOMS project, the committee recommends that the publicly available NAOMS data should not be used for generating rates or trends in rates of safety-related events in the National Airspace System. The data could, however, be useful in developing a set of lessons learned from the project.

1

Introduction and Overview

1.1 ORIGINS OF THE NAOMS SURVEY

NASA's NAOMS project was a survey administered to pilots from April 2001 through December 2004. The origins of NAOMS can be traced to the White House Commission on Aviation Safety and Security, commonly referred to as the Gore Commission. That commission was created to "study matters involving aviation safety and security, including air traffic control and to develop a strategy to improve aviation safety and security, both domestically and internationally."[1] In its report, the commission touched on almost all aspects of the aviation industry and made 57 recommendations in the areas of improving aviation safety, making air traffic control safer and more efficient, improving security for travelers, and responding to aviation disasters. Several of these recommendations can be linked to the NAOMS project, but recommendation 1.1 is of particular relevance:

> Government and industry should establish a national goal to reduce the aviation fatal accident rate by a factor of five within ten years and conduct safety research to support that goal.[2]

This recommendation set in motion a process to create methods for monitoring the National Airspace System (NAS). Existing aviation-data-collection tools for the NAS were limited in scope and did not address all of the areas for which monitoring was believed to be useful.

Thus, the NAOMS survey was developed with the expectation that it would be a new tool within the aviation safety field—a tool that could generate statistically valid rates of events (such as bird strikes or rejected takeoffs) and track trends over time for the entire NAS.[3] The NAOMS survey was not intended to provide an absolute measure of aviation safety;[4] it was intended to support aviation safety policy "in conjunction with the many other data resources available to decision-makers."[5] The use of sample surveys is new in the field of aviation safety, so NAOMS was also viewed as a research study to assess the usefulness of sample surveys in this context.

---

[1] White House Commission on Aviation Safety and Security, *Final Report to President Clinton*, Al Gore, chairman, Washington, D.C., February 12, 1997.

[2] Ibid., p. 9.

[3] Linda Connell, presentation to the Workshop on the Concept of the National Aviation Operations Monitoring Service (NAOMS), Washington, D.C., May 11, 1999, pp. 6-9.

[4] Loren Rosenthal, Manager of NAOMS, Battelle Memorial Institute, "NAOMS Program Overview," presentation to the NRC Committee on NASA's National Aviation Operations Monitoring Service (NAOMS) Project, Washington, D.C., June 9, 2008, p. 5.

[5] Battelle Memorial Institute, *NAOMS Reference Report: Concepts, Methods, and Development Roadmap,* Battelle Memorial Institute, Columbus, Ohio, November 30, 2007, p. 6.

The NAOMS survey was originally designed to collect information on safety-related events as experienced by all of the frontline operators of the NAS, including pilots, controllers, mechanics, flight attendants, and others. However, the survey of pilots appears to have taken longer and required more resources than expected,[6] and surveys of the other groups did not even reach the stage of development. The AC and GA pilots were surveyed from April 2001 through December 2004. NASA decided to discontinue support for NAOMS toward the end of this period, and the survey was transformed into a Web-based tool and handed off to the Air Line Pilots Association (ALPA).[7]

1.2 BRIEF OVERVIEW OF THE SURVEY

The NAOMS project was jointly managed by NASA and the Battelle Memorial Institute, NASA's subcontractor for the project. A total of 29,882 pilots were surveyed as part of the study over the period from April 2001 to December 2004.[8] Of these pilots, 25,105 participated in the AC survey. The survey of GA pilots was conducted only for a brief period (approximately 9 months) and involved 4,777 completed interviews. The FAA's Airmen Certification Database (ACD) was the source from which the sample of pilots to be surveyed was selected.

Each pilot who responded to the survey was asked a set of questions relating to the following: background information, the number of hours and flights that the pilot had flown, the number of events from among numerous possible safety-related events that the pilot had observed, some topic-specific questions, and feedback about the survey. All questions related to events that had occurred within a specific time range (recall period). For most but not all of NAOMS, this recall period was 60 days. Both AC and GA versions of the survey had the same basic structure, with four sections (see Appendixes G and H in this report).

Participation in the NAOMS project was completely voluntary. All data provided by NAOMS respondents were held in confidence. NAOMS maintained records of survey participants, but there is no linkage in NAOMS data repositories between the names of past respondents and the data that they provided.[9] To maintain the confidentiality of the survey participants, NASA released only redacted versions of the survey data. Only these redacted data sets were made available to the NRC's Committee on NASA's National Aviation Operations Monitoring Service (NAOMS) Project: An Independent Assessment.

More detailed descriptions of the sampling design, the survey questionnaires, the redacted data, and other features of the NAOMS project are provided in the following chapters.

1.3 OVERVIEW OF THE REPORT AND SUMMARY OF THE FINDINGS

The rest of this report is organized in six chapters that address the various charges in the committee's statement of task (presented in the Preface).

Chapter 2 discusses ways of measuring aviation safety and describes sources of data on aviation safety that existed before NAOMS or that became available after NAOMS was developed. While fatalities and accident rates are the ultimate measures of aviation safety, the committee notes that there is considerable value in collecting and examining information from certain types of operational data. There were few sources of aviation data when NAOMS was conceived, and many more have become available since then. However, each of these data sources has its advantages and limitations. The FAA's Aviation Safety Information Analysis and Sharing System (ASIAS) is intended to allow easy access to a wide variety of existing databases. It is being developed as a source of data for the entire aviation system.

Chapter 3 assesses the usefulness of sample surveys in providing information on aviation safety. A sample survey is a scientifically valid and effective way to collect data and track trends about events that are potentially related to aviation safety. It can be used to collect reliable information about all segments of civilian aviation,

[6] See NASA, *Final Memorandum on the Review of the National Aviation Operations Monitoring Service,* NASA, Washington, D.C., March 31, 2008, p. 3 (also known as the NASA Inspector General's Report).

[7] Ibid., p. 38.

[8] Battelle, *NAOMS Reference Report,* 2007, p. 13. Other sources provide slightly different numbers, in part because of reclassifications, different data releases, and so on.

[9] Ibid., p. 8.

especially to characterize the safety of general aviation flights and those of other segments of aviation. Further, government-sponsored surveys can provide data that are accessible to the public and can be analyzed independently. However, past experience in the government sector indicates that successful large-scale surveys require a substantial commitment of time and resources to develop, refine, and improve the survey methodology and to ensure that the survey provides useful and high-quality data.

Chapter 4 provides an assessment of the NAOMS sample design. Several aspects of the design, such as the use of a cross-sectional design, the computer-assisted telephone interview (CATI) method, and professionally trained interviewers, are consistent with generally accepted practices in survey design. A CATI system has the potential to incorporate checks for unlikely or implausible values during the interview process. However, the committee found that substantial fractions of responses had implausibly large values for reported hours and flight legs flown as well as event counts, suggesting that the NAOMS survey did not take full advantage of this feature of CATI. The NAOMS team also faced substantial challenges in the choice of the sampling frame and had to make compromises at several stages. Unfortunately, these compromises appear to have led to biases in the sample. While the choices and compromises may have been made for good reasons, the NAOMS team should have investigated the potential impact as well as the magnitude of biases resulting from failure to locate sampled pilots and other forms of nonresponse. In the committee's view, the collection and analysis of supplemental data during the early phase of the survey would have provided a reliable assessment of the various biases and may have led, if necessary, to the development of alternative design strategies.

In Chapter 5, the committee's analysis of the survey questionnaires identified four types of problems that reduced the usefulness of the data collected: (1) the questions went beyond the scope of the intended AC and GA operations, resulting in the aggregation of data from markedly different segments of the aviation industry; (2) some of the questions asked pilots for information that they would likely not have had without a post-flight analysis (for example, the origin of smoke in the aircraft or verification of an uncommanded control surface movement); (3) some of the questions had vague or ambiguous definitions of what constituted an event to be measured; and (4) some of the questions did not have a clear link between the measured event and aviation safety.

As noted above, the committee had access only to the redacted data that were made available to the public. NASA released redacted versions of the survey data starting in December 2007. Chapter 6 describes these redactions and discusses how they further constrain the usefulness of the data and the ability to conduct an analysis to meet the study objectives. One important problem is the grouping of the survey data into whole years. This limits the ability to track changes in event rates over shorter timescales, to determine the effects of changes in the aviation system on event rates, and to assess seasonal and similar types of effects. Issues associated with preserving respondents' anonymity and confidentiality and with the public release of data are common to other government surveys. Such issues should have been anticipated and addressed at the design stage of the project, avoiding the need for after-the-fact, ad hoc redaction methods and the resulting loss of information.

Chapter 7 discusses the potential utility of the NAOMS data. There are several problems with the quality of these data: substantial fractions of responses had implausibly large values, and respondents often rounded their data to convenient numbers. The extent and magnitude of these problems raise serious concerns about the accuracy and reliability of the data. Further, the committee's limited comparison of NAOMS data with other sources indicates that there was an over-representation of some groups and an under-representation of others in the NAOMS survey. Such sampling biases must be addressed in the estimation of event rates and trends. There are many approaches in the survey sampling literature for addressing such biases, but they require detailed information on how the survey was implemented, including the type and nature of problems that were experienced, and access to the original data. In the committee's view, some of these problems could have been reduced substantially if more effort had been spent on ensuring data accuracy during the interview and data-entry stages and if respondents had been asked to refer to their logbooks when possible. The committee does note that many of the biases that relate to the estimation of event rates would be mitigated for trend analysis to the extent that the biases remain relatively constant over time. However, the degree of mitigation might vary substantially across event types.

The committee did not find any evidence that the NAOMS team had developed or documented data analysis plans or conducted preliminary analyses as initial data became available in order to identify early problems and refine the survey methodology. These activities are part of a well-designed survey, especially one conducted as a research study to assess the feasibility of survey methodology in aviation safety.

Based on the analyses and findings of the committee, the publicly available NAOMS data should not be used for generating rates or trends in rates of safety-related events in the National Airspace System. The data could, however, be used in developing a set of lessons learned from the project.

The committee encountered several challenges in assessing NAOMS survey methodology and the potential utility of the data. The committee did not have access to the original, unredacted data. Assessing the utility of the NAOMS data based on heavily redacted data is not an easy task, and it is further complicated by NASA's release of multiple data sets under different redaction methods. Further, as pointed out by the Government Accountability Office's review of NAOMS: "The project staff and contractors did not assemble a coordinated, clear history detailing the project's management that would facilitate evaluation of the overall air carrier pilot survey."[10] The lack of documentation and the delays in obtaining some documents made it difficult for the committee to obtain a full understanding of the steps taken and decisions made (including their rationale) during the NAOMS project. The committee frequently had to rely on statements based on the memory of those involved in the project and, therefore, it received a variety of recollections. Perhaps if decisions had been more clearly documented, there would have been fewer divergent views regarding the various decisions and processes that occurred during the project.

---

[10] Government Accountability Office, *Aviation Safety: NASA's National Aviation Operations Monitoring Service Project Was Designed Appropriately, But Sampling and Other Issues Complicate Data Analysis,* Report to Congressional Requesters, GAO-09-112, Washington, D.C., March 2009, pp. 34-35, 54.

2

Measures of Aviation Safety and Sources of Data

This chapter discusses measures of aviation safety and provides an overview of the various aviation data sources and data that have been available before and since NASA's NAOMS project was developed.

2.1 MEASURING AVIATION SAFETY

How should the safety of the aviation system be measured? The primary goal of most aviation safety initiatives is to reduce the rate of fatalities during or as a result of air travel. Thus, the most important metrics are the rates of fatal accidents per unit of air travel (number of passenger trips, flight legs, flight miles, and so on). For example, the Gore Commission recommended that "the principal focus should be on reducing the rate of accidents by a factor of five within a decade. . . ."[1] Annual reports published by many aviation safety organizations also focus on accident rates and their trends, measured in various ways, and many other studies on aviation safety emphasize fatality and fatal accident rates.[2] For example, the FAA published the Aviation Safety Action Plan report entitled *Zero Accidents: A Shared Responsibility,* in February 1995.[3] The FAA's Safer Skies initiative (April 1998) called for an 80 percent reduction in the rate of fatal accidents by 2007.[4] In the various rate calculations, the denominators (unit of exposure) for accident and fatality rates typically have been the number of aircraft departures, the number of passenger departures, or the number of flight hours.[5]

Fatality rates and fatal accident rates are the most important long-term measures of aviation safety. However, fatal accidents are rare, especially in travel on the major passenger airlines, so looking for changes in fatality rates or fatal accident rates may not give timely feedback regarding the impacts of new equipment, new programs, or changes in the airspace system on the potential risk of air travel.

---

[1] Gore Commission, *Final Report,* 1997, p. 4.

[2] *Fatality rates* measure the rate at which people are killed in aviation and are typically expressed in units of passenger fatalities per million enplanements. *Fatal accident rates* measure the rate at which aircraft are involved in accidents that produce fatalities and are typically expressed in units of fatal accidents per 100,000 flights.

[3] Federal Aviation Administration (FAA), *Zero Accidents: A Shared Responsibility,* Washington, D.C., February 1995.

[4] FAA, Safer Skies: General Aviation Joint Steering Committee, Washington, D.C., June 9, 2009, available at http://www.faa.gov/about/initiatives/safer_skies/, accessed July 15, 2009.

[5] For a discussion of aviation safety measures, see Clinton V. Oster, Jr., John S. Strong, and C. Kurt Zorn, *Why Airplanes Crash: Aviation Safety in a Changing World,* Oxford University Press, Oxford, United Kingdom, 1992, Appendix A.

In an attempt to find a more timely indication of the effects of programs in improving safety or the impact of developments that might lead to a degradation of safety, analysts have turned to other events that they hope might elucidate changes in aviation safety. One approach has been to monitor and study incidents—events involving damage to an aircraft and/or injury, but in which the levels of damage or injury do not meet the NTSB thresholds that define an accident[6]—in the hope of more quickly detecting changes that could potentially affect safety. For example, the FAA Accident/Incident Data System (AIDS) contains reports of collisions between aircraft and birds while flights are on approach to or departure from an airport. Most of these collisions have not resulted in sufficient aircraft damage to be considered an accident by the NTSB. Nonetheless, the rate at which such collisions have occurred can be valuable safety information that might reveal an increase in the presence of birds around airports and suggest potential value in establishing programs to deter birds from nesting in areas adjacent to airports. One important problem, however, is that none of these other indices or events that analysts have turned to have been proven to be precursors to accidents or indicators of a pending increase in fatality rates or fatal accident rates. Even changes in the rates of nonfatal accidents or incidents have not been shown to be predictive of or associated with changes in the rates of fatal accidents. Thus, a link between any of these other indicators and the safety of the aviation system is at best uncertain.

Another complicating factor in understanding and improving aviation safety is that accidents rarely have a single cause.[7] Rather, they are typically the result of a sequence of events involving several malfunctions and/or mistakes. Consider a situation in which an event initiates a sequence of other events that result in damage to an aircraft and/or injury. The same event might occur in another flight but not lead to an accident or injury because something else was done to interrupt the sequence of events. By studying the circumstances and learning how the sequence of events was interrupted, it might be possible to incorporate such information into training or aircraft design and to reduce rates of future accidents. Consider, for example, a flight in which an engine fails during takeoff. With a modern passenger jet aircraft, an engine failure during takeoff should not result in an accident if the pilot takes the correct actions quickly. However, an engine failure places considerable pressure on the pilot, who may not react quickly enough to assess the situation and take the correct action. If one could identify, study, and learn from flights during which actions were taken in time to prevent an accident, it might be possible to improve pilot training so as to reduce the chances of such failures resulting in accidents in the future.

2.2 AVAILABILITY AND SOURCES OF AVIATION DATA

The accident and incident data available from NTSB and FAA are not the only sources of safety data. Two other sources on reports of potentially unsafe events are the Aviation Safety Reporting System (ASRS) and airlines' Aviation Safety Action Programs (ASAPs). ASRS receives, processes, and analyzes voluntarily submitted incident reports from pilots, air traffic controllers, dispatchers, flight attendants, maintenance technicians, and others. ASRS grew out of the FAA's Aviation Safety Reporting Program (ASRP), started on April 30, 1975. The FAA determined that the effectiveness of ASRP would be enhanced if the receipt, processing, and analysis of raw data were done by NASA—an independent third party with no enforcement responsibility—rather than by the FAA. This practice would ensure the anonymity of all parties involved, including the reporter, and would increase the flow of information. Accordingly, NASA designed and administered ASRS through a Memorandum of Agreement executed by the FAA and NASA on August 15, 1975, and subsequently renewed periodically.[8] In 1996, ASAPs were introduced in the flight domain with the hope of encouraging pilots to disclose their errors and, more importantly, the factors

[6] Following is the definition of *incident* from International Civil Aviation Organization (ICAO) Annex 13 and FAA Order 8020.11b: "an occurrence other than an accident associated with the operation of an aircraft, which affects or could affect the safety of operations." For more information on thresholds between an accident and an incident, see National Transportation Safety Board (NTSB), *Form 6120.1 Pilot/ Operator Aircraft Accident/Incident Report,* NTSB, October 2006, available at http://www.ntsb.gov/aviation/6120_1_printonly.pdf, accessed June 11, 2009.

[7] James Reason, *Human Error,* Cambridge University Press, Cambridge, United Kingdom, 1990.

[8] See FAA, Advisory Circular 00-46D, Washington, D.C., February 26, 1997, available at http://rgl.faa.gov/Regulatory_and_Guidance_Library/rgAdvisoryCircular.nsf/8e17c23e2f26e8018625726d006ce776/64358057433fe192862569e7006da716/$FILE/AC00-46D.pdf, accessed July 15, 2009.

contributing to their errors.[9] Both ASAPs and ASRS allow for voluntary reports initiated by pilots and others in the aviation system, and both systems provide some degree of immunity from regulatory action or civil fines. These reports can play a valuable role in understanding and improving aviation safety. However, because they are based on voluntary reporting (resulting in a nonprobability sample; see Chapter 3 for a discussion of types of samples), they do not provide a means to develop statistically valid metrics of aviation safety.

Around the same time that NAOMS was developed, the Flight Operational Quality Assurance (FOQA) program was started by several airlines in the mid- to late 1990s. Through their FOQA programs, airlines gather selective digital data from flight data recorders installed on their airplanes and then process and analyze those data to identify particular events. These data can provide extensive information about flight operations. Unlike ASRS and ASAP reports, FOQA data can be available from all flights on all properly equipped airplanes. Unfortunately, these data are not currently available for all segments of the NAS. Flight data recorders are typically not installed on general aviation airplanes, and some operators who do have airplanes with appropriate flight data recorders may not have the resources to gather, process, and analyze those data. However, both recent and future advances in computer and data-storage technology may well reduce the cost of both collecting and analyzing such digital data. It may also become easier to integrate FOQA data with other data sources such as air traffic control data from FAA's Operational Error Detection Program. FOQA data are limited, however, in that they do not reveal the intentions of the pilot during the recorded events, so they may provide an incomplete picture of the event.

More recently, FAA has launched the Aviation Safety Information Analysis and Sharing System (ASIAS). This system is housed and managed at the MITRE Corporation and contains both aviation safety and operations data and a collection of studies of specific aviation safety topics. ASIAS is designed to enable users to perform integrated inquiries across multiple databases, search an extensive warehouse of safety data, and display pertinent elements in an array of useful formats. ASIAS contains both the data sets and the query tools that allow easy access to the data.[10]

ASIAS is being developed in a phased approach. One can already access an array of aviation safety databases, including FAA Accident/Incident Data Systems, the Air Registry, the ASRS, Bureau of Transportation Statistics, the Near Midair Collision System, the NTSB Aviation Accident and Incident Data System, NTSB Safety Recommendations to the FAA with FAA Responses, and the World Aviation Accident Summary.[11] Additional databases planned for inclusion in ASIAS include ASAP data, FOQA data, and other data.[12] (A variety of other data sets are also available to the FAA and to safety analysts that are not included or planned to be included in ASIAS.)

The use of statistical techniques to extract pertinent information from currently available data is an attractive approach, as it takes advantage of information that is already available. However, extracting, combining, analyzing, and understanding data from diverse sources involves many challenges. MITRE and others are working to develop additional tools to extract information from ASIAS data.

All of the currently available data sources have their advantages and limitations. ASIAS has made considerable progress in allowing a wide variety of these databases to be easily accessed and integrated. If ASIAS continues to be developed as planned, it can become an even more useful source of data for the entire aviation system.

---

[9] See FAA, Advisory Circular 00-58, Washington, D.C., May 4, 1998, available at http://rgl.faa.gov/Regulatory_and_Guidance_Library%5CrgAdvisoryCircular.nsf/list/AC%2000-58/$FILE/AC00-58.pdf, accessed July 15, 2008.

[10] FAA, *Aviation Safety Information Analysis and Sharing (ASIAS) System,* Washington, D.C., available at http://www.asias.faa.gov/portal/page?_pageid=56,398034,56_398041&_dad=portal&_schema=PORTAL, accessed May 16, 2009.

[11] Ibid.

[12] FAA, *National Aviation Research Plan (NARP) Appendices,* Washington, D.C., February 4, 2008, available at http://www.faa.gov/about/office_org/headquarters_offices/ato/service_units/nextgen/research_planning/narp/media/pdf/NARP_08_APP.pdf, accessed July 15, 2009; and Victoria Cox, Statement before the House Committee on Transportation and Infrastructure, Subcommittee on Aviation, March 18, 2009, available at http://www.faa.gov/news/testimony/news_story.cfm?newsId=10433, accessed July 15, 2009.

3

Sample Surveys: Overview, Examples, and Usefulness in Studying Aviation Safety

Task 2 of the charge to the committee is to "assess the advantages and disadvantages of using a survey method to collect . . . a statistically meaningful data set" to estimate and characterize the safety of the NAS. This chapter begins with an overview of sample surveys, which are not widely used in the aviation community. Other federal agencies have been using sample surveys successfully for a long time. Section 3.2 discusses some of these surveys, including a detailed description of the National Crime Victimization Survey (NCVS), which shares some key features with the NAOMS survey. A few additional examples of government surveys are given in Appendix E. Section 3.3 discusses the usefulness and limitations of sample surveys for collecting aviation safety data.

3.1 OVERVIEW OF SAMPLE SURVEYS

In statistical terminology, a *sample* refers to a subset of the population of interest. Note that many of the aviation data sets discussed in Section 2.1 are samples, because the available data are a subset of the data sets for the whole aviation system.

Samples can be grouped broadly in two categories: probability samples and nonprobability samples. In probability sampling, the subset is selected according to a specified probability mechanism. This provides a basis for using sample data to draw appropriate statistical inference (point and interval estimates, statements about statistical bias and precision, and so on) about the population characteristic(s) of interest. The uncertainty in the estimate because of sampling variability is referred to as the *sampling error* or *margin of error*. Nonprobability sampling (such as judgment or convenience sampling techniques) does not allow one to make a similar inference about the population characteristics without additional assumptions.

The term *survey* refers to techniques for collecting data from the target population of interest. While surveys are generally identified with human populations (for example, opinion polls, consumer surveys, demographic and economic surveys), surveys of other types of populations (such as geological surveys and administrative records) are also common. A survey that collects data from the entire population is called a *census*. In most situations, however, data are collected from only a subset of the population, in which case the survey is called a *sample survey*. As noted above, one must use probability sampling methods in order to make statistically valid conclusions about the target population.

There are many ways of selecting probability samples, but the simplest method is simple random sampling. Under this method, every possible subset of a fixed size from the population has equal probability of being selected. For practical and statistical reasons, it may not always be desirable or feasible to use simple random sampling. A variety of other probability sampling techniques, such as stratified sampling, cluster sampling, and multistage sampling, as well as corresponding estimation methods, have been developed in the literature.[1]

In surveys of human populations, the data are generally collected using a questionnaire as the survey instrument—the participants are asked to respond to a set of questions. The design of the questionnaire is critical to ensuring that the data collected are of good quality and can provide information that is generalizable to the target population. There is a vast literature on questionnaire design.[2] There are also many ways of conducting surveys of human populations—for example, by mail, through telephone interviews, or in on-line surveys.

Since a sample survey collects data from only a subset of the population, the estimates have sampling error. The use of probability sampling methods allows one to characterize and estimate this error. The sampling error depends on the probability sampling method used; methods for estimating the sampling error are discussed extensively in the literature.[3]

There are also several other types of errors (often called nonsampling errors) that occur commonly in surveys (including censuses). For example, *coverage bias* can occur when the *sampling frame* (a list of identifiable units from which to draw the actual sample, such as identification numbers, geographical coordinates, household addresses, or telephone numbers) is incomplete. The sampling frame may systematically miss some classes of population members entirely. The sampling frame may also include units that are not members of the target population. Also, failures to contact sampled subjects or nonresponse by those who are contacted can lead to additional biases. *Measurement error* can result when the survey instrument is poorly designed or if problems arise in the field implementation of the survey.

The nature and the magnitude of added uncertainty because of nonsampling errors cannot be ascertained from the sample itself, regardless of whether it is a probability or nonprobability sample, or even if it is a census. Thus, an important part of the survey planning and implementation process is to determine ways to make these errors as small as possible.[4]

There are several steps in planning and implementing a good survey. For the purpose of the discussion here, the relevant steps include the following:

- Identify the population of interest (the set of units from which the survey would ideally collect data in the absence of concerns over cost or respondent burden) and the characteristic(s) to be studied.
- Determine the method(s) for conducting the survey (such as mail or telephone interviews) and implementing the survey.
- Develop a sampling frame that will be used to select the sample.
- Determine a sampling design, the probability sampling method and the sample size, and the number of elements to be selected (the latter depends on the sampling design and the desired precision as well as on available resources).
- Design the data-collection instrument (for example, the questionnaire for a human population).
- Examine possible sources of error, ways to reduce them, and ways to estimate them.
- Analyze the data and report the results.

Despite the presence of sampling errors, sample surveys have several advantages over censuses:

- Samples are less costly than censuses. For example, for populations with large, hard-to-find, or highly

[1] William G. Cochran, *Sampling Techniques,* Wiley, New York, 1977; and Sharon Lohr, *Sampling: Design and Analysis,* Duxbury Press, Pacific Grove, Calif., 1999, pp. 4-8.

[2] Norman M. Bradburn, Seymour Sudman, and Brian Wansink, *Asking Questions: The Definitive Guide to Questionnaire Design,* Jossey-Bass, San Francisco, Calif., 2004.

[3] Cochran, *Sampling Techniques,* 1977; and Lohr, *Sampling: Design and Analysis,* 1999.

[4] Judith Lessler and William Kalsbeek, *Nonsampling Error in Surveys,* Wiley, New York, 1992.

dispersed (in space or time) units, the cost of locating and collecting data from the units can be high. Censuses also require more resources to train the data-collection staff.

- Data from sample surveys can be collected and analyzed in a more timely manner than can data from censuses.
- It is difficult to control nonsampling errors in large or difficult-to-reach populations. For example, collecting data from all the units in these populations would require a large and dispersed administrative staff, which would be harder to train and closely supervise. Nonresponse problems are also harder to manage in a larger operation.
- If a census is repeated over time, it would require all the units in the population to be repeatedly monitored. For human populations, this would place considerable burden on the respondents and could lead to decreased cooperation and higher nonresponse rates.

3.2 THE USE OF SAMPLE SURVEYS IN THE GOVERNMENT SECTOR

Sample surveys are used routinely by countries around the world to collect and analyze data in order to inform policy decisions, allocate resources, and assess national needs. Various federal agencies in the United States (for example, the Census Bureau, the Bureau of Labor Statistics, the Bureau of Justice Statistics) have been conducting or sponsoring sample surveys to obtain high-quality data about the state of the economy, health, education, crime, and other issues. The U.S. decennial census is mandated by the U.S. Constitution for the purpose of allocating congressional seats, but virtually all other demographic information used for policy making is collected on a sample basis. Several other countries have replaced their censuses with sample surveys (for example, Germany in 1987 and France in 2004). Even in the United States, the long form for the census was replaced in 1996 by the American Community Survey, a monthly sample survey of about 250,000 households. [5]

The rest of this section describes one particular survey, the National Crime Victimization Survey (NCVS), in some detail as it shares some key similarities with the NAOMS survey: it uses multiple data sources, many of which are self-reported; it is a national survey designed to collect sensitive data (crime versus aviation safety); and it must protect respondent confidentiality. It is also informative to see how this survey started and how it evolved over time. See Appendix E for additional examples of federal surveys.

Several sources of crime data are used to inform U.S. policy decisions and to allocate funding for criminal justice to the states. The Uniform Crime Report (UCR) began in the late 1920s when the International Association of Chiefs of Police recognized a need for reliable data on crime in the United States in order to measure the effectiveness of local law enforcement and to provide data to help fight crime. In 1930, the job of collecting, summarizing, and publishing the UCR was turned over to the Federal Bureau of Investigation (FBI), which received data on monthly counts of eight types of crimes as well as the number of arrests for an additional 21 crimes from police jurisdictions. Although participation in the UCR is voluntary, over 98 percent of all police agencies in the nation reported to the UCR for at least some months in 2005.[6] However, the system had some weaknesses in that some police agencies reported on a wide range of characteristics of all crimes, whereas others submitted information on a more limited set of crimes.

In the 1970s, the criminal justice community determined a need for more in-depth information about reported crime incidents. The Bureau of Justice Statistics commissioned a study to determine how the UCR could be improved to meet these needs. Based on that study, the UCR was further refined into what is now known as the National Incident Based Reporting System (NIBRS). By 2007, only about 25 percent of the U.S. population was covered by a police jurisdiction reporting to the NIBRS,[7] which has not yet replaced the UCR for national crime statistics. However, even if the NIBRS included data from all jurisdictions, there would still be gaps in the available information about crime because not all crime is reported to the police.

---

[5] U.S. Census Bureau, *American Community Survey (ACS),* Washington, D.C., available at http://www.census.gov/acs/, accessed July 15, 2009.

[6] Nathan James and Logan Richard Council, *How Crime in the United States Is Measured,* CRS-RL34309, Congressional Research Service, Washington, D.C., 2008.

[7] Federal Bureau of Investigation, *NIBRS Frequently Asked Questions,* Washington, D.C., April 2009, available at http://www.fbi.gov/ucr/nibrs_general.html, accessed October 21, 2009.

Other developments were occurring in parallel. Several years of pilot tests showed that the URC seriously underestimated the level of crime in the United States and that the collection of data from the victims of crime was feasible. In 1965, President Lyndon Johnson established a commission to examine the data needs with respect to crime statistics and to propose a solution. The commission recommended in 1968 that a Justice Statistics Center be established and that a national crime survey be implemented on an ongoing basis. A multiyear period of research and experimentation was conducted by the Census Bureau, which was selected to implement the survey. The first National Crime Survey, or NCS, was conducted in July 1972. An NRC panel was asked to examine the NCS because of concerns about its data.[8] Based on the recommendations in the panel's 1976 report, *Surveying Crime,* the Census Bureau sharpened some questions to better define certain types of incidents for respondents and to collect additional data allowing better comparisons with other data sources. These changes were phased in from July 1986 through 1992, and the NCS was renamed the National Crime Victimization Survey. For 18 months, both surveys were administered, each to half of the sample, so that comparisons could be made between them and methods for bridging the time series could be developed. In 2005, the NCVS interviewed people in a sample of about 68,000 households.

These two sources of crime data—the UCR/NIBRS and the NCVS—provide complementary information to policy makers. The NCVS provides coverage for the large number of crime incidents not reported to the police. Even for crimes that are reported, the NCVS collects information known only to the victim, such as the impact of the crime on his or her life. The Department of Justice says of the two systems that "the information they produce together provides a more comprehensive panorama of the Nation's crime problem than either could produce alone."[9]

Comparison of crime rates from the two crime data-collection systems is inevitable. Some serious crime categories have high reporting rates, and for these, the magnitudes of criminal incidents obtained from the UCR/NIBRS and the NCVS are similar. For other crime categories, the counts can be quite different. These differences have been investigated by many researchers and can be explained by the different methodologies, definitions, and error types of the two collection systems.[10]

Victims of crime may be reluctant to reveal details of a crime for fear of embarrassment or compromise to their safety. To encourage participation in the NCVS, victims are provided assurances of confidentiality. These promises are enforced through a variety of federal regulations, which provide restrictions on how the information victims reveal can be used (only for statistical purposes) and with whom it can be shared (the data are immune from legal processes). A program allowing researchers outside the federal data-collecting agency access to these data requires that they obtain a privacy certificate verifying the security of their data-management plan.[11]

The NCVS is a useful and high-quality data-collection system for the geographically dispersed, sensitive, and diverse phenomenon of crime incidents. While this type of system can be costly, policy makers have deemed it valuable enough to justify the expense. It did not achieve its maximum utility immediately at its inception, but rather required adaptation and improvement over time. The NCVS example also illustrates some features that are relevant for NAOMS: (1) data available from a system of self-reports are not necessarily adequate in characterizing the complete picture on crime; (2) when two systems produce different estimates for certain categories of crimes, it does not necessarily invalidate the utility of either; the differences can provide insight into how to improve measurement and how to determine the most useful concepts or definitions being examined; and (3) confidentiality

[8] National Research Council, *Surveying Crime,* Panel for the Evaluation of Crime Surveys, Bettye K. Eidson Penick, editor, and Maurice E.B. Owens III, associate editor, Committee on National Statistics, Academy of Mathematical and Physical Sciences, National Academy of Sciences, Washington, D.C., 1976.

[9] U.S. Department of Justice, *The Nation's Two Crime Measures,* U.S. Department of Justice, Washington, D.C., October 2004, p. 1, available at http://www.ojp.usdoj.gov/bjs/pub/pdf/ntcm.pdf, accessed June 10, 2009.

[10] Michael R. Rand and Callie M. Rennison, "True Crime Stories? Accounting for Differences in Our National Crime Indicators," *Chance* 15 (2002): 47-51, available at http://www.ojp.usdoj.gov/bjs/pub/pdf/tcsadnci.pdf, accessed June 12, 2009; and James P. Lynch and Lynn A. Addington, eds., *Understanding Crime Statistics: Revisiting the Divergence of the NCVS and UCR,* Cambridge University Press, New York, 2007.

[11] See Bureau of Justice Statistics, *Protection of Human Subjects and Privacy Certificate Requirements for Applicants for Funding from the Bureau of Justice Statistics,* U.S. Department of Justice, available at http://www.ojp.gov/bjs/pub/pdf/bjshs.pdf, accessed March 19, 2009.

protections can be put in place by regulation so that the respondents can be protected at the same time that the data are serving useful purposes for policy making and research.

In summary, most large-scale surveys evolve over time, and their survey methodologies are refined on the basis of experience before they attain excellence. Often, the changes are gradual, but some surveys have undergone major design changes. Examples include the NCVS as well as the National Assessment of Educational Progress. Another feature of successful government surveys is that they typically have a research team and resources to support the investigation of issues or problems of particular import. They also have a core staff dedicated to the survey's ongoing improvement and adaptation to change. Many large-scale survey programs also develop an organizational culture that fosters a professional approach to the development of survey methodology and produces staff in both technical and administrative areas that are very knowledgeable about a particular survey, its history, and its key issues.

Finding: Successful large-scale surveys typically require a substantial commitment of time and resources to develop, refine, and improve the survey methodology and to ensure that the survey provides useful and high-quality data.

3.3 USEFULNESS OF SAMPLE SURVEYS FOR ASSESSING AVIATION SAFETY

When NAOMS was proposed, the available sources of data on aviation safety included the following: (1) accident and incident data from the NTSB and the FAA, (2) data from the FAA's NASA-operated ASRS, (3) FAA's Near Midair Collision Database, and (4) FAA's Operational Error Detection Program. As noted in Section 2.1, the NTSB and FAA accident and incident databases include only incidents that meet certain thresholds and so do not include all potentially unsafe occurrences. The ASRS database consists largely of self-reports of incidents by pilots. Though it is large and rich in information, it is not a probability sample, so it is impossible to obtain statistically valid estimates from ASRS data. The same limitation holds for the Near Midair Collision Database. The Operational Error Detection Program data cover only aircraft operating in controlled airspace, so planes flying under visual flight rules, which include a high proportion of general aviation flights, would not be covered.

In recent years, the use of onboard data-acquisition systems to collect aircraft operations data is becoming common. Now, virtually all new commercial airliners and most high-end business jets are equipped with flight data recorders, which provide the basis for FOQA systems that provide detailed information about flight operations. These systems are not affected by the types of measurement errors that are present in surveys of pilots or other personnel. However, as noted in Section 2.1, FOQA data do not provide a complete picture of the entire airspace, as piston-engine and turboprop general aviation aircraft are not typically equipped with these data-collection systems. Even if it were possible to obtain FOQA data for the entire population of aircraft in the U.S. airspace, the resources involved in assembling the data from all the air carriers and in ensuring privacy and confidentiality so that the data could be shared among all the carriers, the government, and the public would be very high. The use of probability sampling techniques can be useful here, as one could collect and analyze a sample of the database that takes into account privacy and confidentiality considerations in order to obtain timely information at reasonable costs.

The Aviation Safety Information Analysis and Sharing System (discussed in Section 2.1) has made progress and shows promise in allowing the access to, analysis of, and integration of multiple large aviation safety data sets. As it continues to develop and as more data sets are added, it will become more comprehensive. However, even then it is unlikely to cover the entire aviation system, particularly general aviation and small commercial carriers operating in remote locations. In addition, as more databases are added, issues of privacy and confidentiality are likely to take on increasing importance before the data can be shared among all parties and with the public.

Sample surveys can be used to provide new or supplemental information about aviation safety, even in the presence of these other data-collection efforts. The scope and usefulness of NAOMS are explored in detail in the next two chapters, but generally speaking, NAOMS was an attempt to capture the experiences of the frontline personnel (pilots, flight attendants, air traffic controllers, and mechanics) regarding flight operations and aviation safety. In the committee's view, such information could be potentially useful, particularly in those segments of aviation that are not well covered by the other databases. In addition, carefully planned surveys can provide useful information not only about specific events, but also about the views and perceptions of the frontline personnel on flight operations. However, care must be taken to solicit information from these frontline personnel only when they are in a position to provide accurate and consistent responses.

Finding: A sample survey is a scientifically valid and effective way to collect data and track trends about events that are potentially related to aviation safety. The sample survey has several advantages over other, currently available, data sources:

- **Sample surveys can be used to collect reliable information about all segments of civilian aviation. They can be especially useful for characterizing the safety of general aviation flights and the safety of flights of other segments of aviation where the data are more limited.**
- **Sample surveys have the potential to generate statistically valid information about operations that may or may not result in an accident or incident. This information would provide a useful reference point for studying other event data and for learning why some events lead to accidents while other, similar events do not.**
- **Government-sponsored sample surveys can produce data that are accessible to the public and can be analyzed regularly and independently.**

However, information from any survey should be used in conjunction with other existing data to provide a holistic assessment of aviation safety levels and trends.

4

Assessment of NAOMS Sampling Design

As noted in Chapter 1, the goals of the NAOMS survey were to estimate event rates and trends in the rates for a variety of safety-related events. This chapter assesses the impacts of the sample design and potential coverage biases on the accuracy of these estimates.

4.1 INTRODUCTION

Research by the NAOMS team indicated that there were more than 600,000 active pilots in 1998, with about 130,000 having Airline Transport Pilot Certificates.[1,2] The team identified two ways to build the sampling frame of pilots for the NAOMS survey: use the FAA's Airmen Certification Database (ACD)[3] or partner with industry trade groups and unions to obtain pilots' contact information. The team eventually decided to use the ACD because it was logistically simpler and did not run the risk of compromising the independence of the survey.

The NAOMS team decided that pilots in the ACD who met four criteria—(1) being based in the United States, (2) having an airline transport pilot (ATP) certificate, (3) having a multi-engine rating, and (4) having a flight engineer (FE) certificate—were eligible for selection for the AC survey. All other pilots were eligible for the GA survey. Budgetary and statistical considerations led the NAOMS team to set a goal of 8,000 completed AC survey questionnaires each year.[4]

The criteria outlined above resulted in a pool of 52,570 pilots for the AC survey.[5] The NAOMS team further narrowed the sample pool by eliminating any pilot who could not be linked to a telephone number.[6] Actual

---

[1] Battelle, *NAOMS Reference Report,* 2007, Appendix 2, p. 2.

[2] For a description of the Airline Transport Pilot Certificate, see Federal Aviation Administration, *Airline Transport Pilot, Aircraft Dispatcher, and Flight Navigator Knowledge Test Guide,* Washington, D.C., September 2008, available at http://www.faa.gov/training_testing/testing/airmen/test_guides/media/FAA-G-8082-1D.pdf, accessed July 15, 2009.

[3] Available at http://www.faa.gov/licenses_certificates/airmen_certification/interactive_airmen_inquiry/, accessed July 15, 2009.

[4] Mary Connors, Chief, Aviation System Safety Research Branch, NASA Ames Research Center, presentation at Meeting 3 of the National Research Council (NRC) Committee on NASA's National Aviation Operations Monitoring Service (NAOMS) Project, NASA Ames Research Center, Moffett Field, California, January 14, 2009, p. 14.

[5] Battelle Memorial Institute, *NAOMS Completion Rate Summary: Air Carrier and General Aviation Surveys,* Columbus, Ohio, August 31, 2008, p. 4.

[6] According to its *NAOMS Completion Rate Summary,* Battelle could not "obtain good telephone numbers" for 9,480 out of 52,570 pilots (3,590 out of 12,363 for the GA survey).

implementation of the survey reached approximately 7,000 air carrier pilots each year.[7] A total of 29,882 pilots were surveyed for the NAOMS study over the period from April 2001 to December 2004.[8] Of these pilots, 25,105 participated in the AC survey. The GA survey was conducted for a much shorter period (August 2002 through April 2003) and involved 4,777 pilots. Groups of pilots for both surveys were selected monthly using a simple random sampling from the public ACD.

The sampled pilots were contacted first by mail with a pre-notification letter from the NAOMS team. This letter was followed by a telephone call during which the survey was administered. If the respondent was not available, a callback time was arranged. The survey questionnaire included a computer screen to allow checking for qualifying activity during the *recall period*, which consisted of the last *n* days before the survey, with the number *n* varying initially from 30 to 90 days but fixed at 60 days after March 2002. The survey was conducted by professionally trained interviewers using a computer-assisted telephone interview system.

Each pilot who responded to the survey was asked a set of questions—described in detail in Chapter 5—about his or her background, the number of hours and flights flown, the number of numerous possible safety-related events observed, some topic-specific questions, and feedback about the survey. The information in the responses was restricted to the recall period.

4.2 TARGET POPULATION AND SAMPLING METHOD

For the NAOMS surveys, the two target populations were all flight legs meeting the criteria for AC and GA during the recall period. The NAOMS questionnaires[9] indicate that the qualifying AC flight legs were intended to be those conducted under Federal Aviation Regulations (FAR) Part 121 (under which the major passenger and large cargo airlines such as FedEx fly).[10] Considering air carrier operations as those operating under Part 121 is consistent with the practices of the U.S. Department of Transportation.[11] The flights of interest in the GA questionnaire were those conducted under FAR Parts 91[12] and 135. However, because FAR Part 135 governs the operation of scheduled commuter air carriers and on-demand, for-hire air taxi and charter providers,[13] the inclusion of flights operated under Part 135 in the general aviation survey extended the notion of general aviation well beyond normal usage of the term. In its general aviation safety statistics, the U.S. Department of Transportation specifically excludes Part 135 operations and considers Part 135 scheduled operations to be a segment of aviation separate from both general aviation and also from Part 135 on-demand operations.[14] The GA survey did not collect the information that would have enabled events from these disparate segments to be disaggregated.

The ideal sampling frame for this population would be the list of all flight legs that occurred in the appropriate flight regimes, that is, Part 121 flights in the AC survey and Part 91 and Part 135 (given the NAOMS definition of general aviation) in the GA survey during the recall period. However, collecting data for a simple random sample of flight legs would not have been economical or even feasible. The NAOMS team decided to draw samples of pilots and to ask them about all events that occurred during the recall period. This strategy results in a cluster sampling of flight legs: pilots are the primary sampling units, and all the flights flown by the sampled pilots during the recall period are then included in the sample.

Such a cluster sample of flights differs from a simple random sample in several ways. In particular, the flight

---

[7] Battelle, *NAOMS Reference Report,* 2007, p. 34.

[8] Ibid., p. 13. Other sources provide slightly different numbers, in part because of reclassifications, different data releases, and so on.

[9] Ibid., Appendixes 11 and 12.

[10] *FAR Part 121* refers to a section of the FAA Federal Aviation Regulations that prescribes safety rules governing the operation of air carriers and commercial operators of large aircraft. The term *Part 121 carriers* refers to carriers operating under these regulations; see Air Transport Association, *The Learning Center: Glossary,* 2009, available at http://learningcenter.airlines.org/Pages/Default.aspx?Filter=p, accessed July 15, 2009.

[11] See, for example, Bureau of Transportation Statistics, *National Transportation Statistics 2009,* Research and Innovative Technology Administration, U.S. Department of Transportation, Washington, D.C., 2009, Table 2-9: U.S. Air Carrier Safety Data.

[12] *FAR Part 91* refers to a section of the FAA Federal Aviation Regulations that includes principally general aviation.

[13] Air Transport Association, *The Learning Center: Glossary,* 2009.

[14] See, for example, Bureau of Transportation Statistics, *National Transportation Statistics 2009,* Table 2-9: U.S. Air Carrier Safety Data.

legs of any particular pilot are either sampled or not sampled as a group. This typically reduces the information content relative to a simple random sample of the same size because the responses within clusters are likely to be correlated.[15] However, it is often much more economical to use cluster sampling, in which case cost reductions lead to greater overall efficiency. This is the case with the NAOMS survey. Cluster sampling occurs in many surveys—for example, samples of students within schools or patients within hospitals.

A second problem arose in the NAOMS survey owing to the fact that there can be multiple pilots on any given flight. This implies that the probability of sampling flight legs varies with the number of pilots on the flight. There is also a non-zero (although small) probability that a flight leg will be selected multiple times in the sample. Estimates of rates of events must account for these unequal sampling probabilities, but this requires knowledge of the probabilities, which is difficult to obtain. Section 7.3 discusses this issue further.

Finding: The decision to use pilots as the sampling units in the NAOMS project was appropriate for efficiency reasons. While this led to a cluster sampling of the basic units of interest (flight legs or hours), the costs involved in sampling flight legs would have been prohibitive. However, since there can be multiple pilots on flights, this scheme results in unequal sampling probabilities for flight legs. While the NAOMS study team was aware of the problem, the team did not examine its extent or consequences and the team did not develop methods to address the problem for estimating rates of events.

There are two issues with the use of the sampling frame in the NAOMS survey. The first is whether the appropriate pilots were sampled, which is discussed in the next section of this chapter. The second issue is whether the selected pilots' flight legs were confined to those that are in the operations of interest, as many pilots fly in more than one type of operation during the recall period. This issue is addressed in Chapter 5.

4.3 COVERAGE ISSUES

4.3.1 Opting Out of the Database

One source of potential problems with the publicly available version of the ACD is that starting in 2000, the FAA allowed pilots to "opt out" of the database. This opt-out option resulted in an incomplete sampling frame. The NAOMS project team considered other options, including the possibility of obtaining pilot names from industry trade groups and/or organized labor. Those options were not adopted because of the challenges of merging multiple lists and because of concerns about limitations that list providers might place on the project.

The committee was able to get data on opt-outs only for 2008, which showed that only 6 percent of all pilots opted out. However, pilots with certificates associated with commercial activity opted out at much higher rates: 20 percent for those with ATP certification and 36 percent for those with FE certification.[16] The apparently large coverage gap for the AC sampling frame raises the potential for substantial bias in observed outcomes if event rates for pilots who opted out differed from those who did not.

The public version of the ACD was also used as the initial sampling frame for the GA survey that was conducted during August 2002 through April 2003. The opt-out provision does not appear to have posed much risk to the GA sampling frame. In 2008, coverage of pilots without an FE certificate (the closest available approximation to the GA sampling frame) was 96 percent.[17]

Finding: If the event rates for pilots who opted out of the public Airmen Certification Database differed considerably from those who did not, the high opt-out rates would have resulted in substantial biases in the AC survey. The use of the ACD does not appear to have been a serious problem for the GA survey.

---

[15] Cochran, *Sampling Techniques,* 1977.

[16] Harold Everett, Manager, Airmen Certification Branch, Civil Aviation Registry, FAA, personal communication to Anthony Broderick, Member, NRC Committee on NASA's National Aviation Operations Monitoring Service (NAOMS) Project: An Independent Assessment, June 13, 2008.

[17] Ibid.

4.3.2 Stratification for Creating the AC and GA Sampling Frames

Most pilots in the public ACD are not commercial pilots. Consequently, a random sample from the ACD would include many pilots who are ineligible for the AC survey. (At the beginning of the NAOMS study, the ACD included about 640,000 pilots, while estimates for the number of AC pilots ranged between 75,000 and 90,000.[18]) Because there was no information in the database to directly identify commercial pilots, the use of the full ACD as a sampling frame for the AC survey would have required contacting and screening at least seven times as many pilots as were desired for the final sample.

The NAOMS team deemed that this screening would be too costly, so it decided instead to filter the ACD for pilots with certifications indicative of pilots who fly for air carriers. Specifically, the AC sampling frame was limited to U.S.-based pilots who had an ATP certificate, multi-engine rating, and an FE certificate. However, some active AC pilots do not meet all of these criteria. Many AC pilots, including captains and first officers, do not hold an FE certificate. In addition, it is not necessary for a first officer to hold an ATP certificate; a commercial certificate is all that is required to be a first officer. Thus, the criteria used in the NAOMS survey meant that some members of the target population were eliminated from the sampling frame.

The GA sampling frame was essentially the complement of the AC frame—that is, it included all of the pilots in the ACD who lacked one of the certificates needed for inclusion in the AC frame.[19] Before starting with the GA questionnaire, pilots were asked about their involvement in various types of flying activity during the preceding 60 days. Pilots with AC activity but no GA activity were administered the AC questionnaire. Pilots with both AC and GA activity were administered either the AC or GA questionnaire at random. In contrast, pilots selected from the AC sampling frame were not screened for administration of the GA questionnaire.

Table 4.1 shows the raw distribution of the aircraft types in the NAOMS survey for the 4-year period (2001 through 2004). Information from the U.S. Department of Transportation and Bureau of Transportation Statistics is also shown for comparison. It is clear that wide-body aircraft are over-represented in the NAOMS survey and that small aircraft are under-represented. There are several possible reasons for these differences. They could be due, at least in part, to the requirement for an FE certificate. In the earlier days of jet travel, the typical cockpit crew consisted of three people (a pilot, a first officer or copilot, and a flight engineer), and it was common to start as a flight engineer, advancing to the first officer and then to the pilot position. More recently, jet aircraft have been designed to eliminate the flight engineer position and require only a pilot and a first officer. As a result, pilots with an FE certificate are more likely to be older and more senior pilots and are also more likely to be flying wide-body aircraft. A second possible reason for the difference in representation of wide-body and small aircraft in the NAOMS survey is the unequal sampling probabilities for the flight legs owing to the differences in the number of pilots in the aircraft, with wide-body aircraft having more pilots than other aircraft. However, in the committee's view, the large disparities in the numbers in Table 4.1 are more likely to have been caused by the FE certificate requirement.

To the extent that these differences led to different event rates, estimates from the AC sample would be biased. Appropriately weighting the results by aircraft or aircraft type (for example, by giving more weight to results from pilots who flew small aircraft) might reduce the size of some of these biases (see the discussion in Chapter 7). But weighting of the data from the main AC sample could not compensate for biases within aircraft type owing to pilots without one of the certificates required for the AC sampling frame.

Just as the AC sampling frame excludes some of the AC target population, the GA sampling frame excludes some of the GA target population. Pilots with the certifications required for the AC sampling frame (and therefore excluded from the GA frame) can and often do participate in general aviation. Because pilots contacted as part of the AC survey were not asked about GA activity, there does not appear to be any way to estimate the size of the coverage gap with the NAOMS data.

Finding: The NAOMS AC sampling frame was restricted to pilots who hold ATP and FE certifications. This restriction excluded many active air carrier pilots and appears to have led to biases such as over-representation of wide-body aircraft and under-representation of small aircraft in the NAOMS sample. The

[18] Connors, presentation to NRC Committee on NAOMS, 2009, p. 13.

[19] Ibid., p. 17.

TABLE 4.1 Proportion of Aircraft by Model Size in NAOMS Survey Compared to Bureau of Transportation Statistics (BTS) Data for the Same Period

	2001		2002		2003		2004		2001-04	
Model Size	NAOMS	BTS	NAOMS	BTS	NAOMS	BTS	NAOMS	BTS	NAOMS	BTS
Wide-body	29.0	17.2	27.8	17.9	29.3	16.9	30.9	16.3	29.2	17.0
Large	17.0	13.7	15.8	14.0	15.4	25.2	16.3	12.0	16.0	16.1
Medium	49.1	52.6	48.6	52.5	48.1	38.9	47.4	48.7	48.3	48.2
Small	4.9	16.5	7.8	15.6	7.2	19.1	5.3	22.9	6.5	18.7

SOURCE: Data gathered from NASA, *NAOMS 2008 Air Carrier Responses by Category*, NASA, Washington, D.C., available at http://www.nasa.gov/news/reports/NAOMS_08_ac_resp_category.html, accessed June 11, 2009; and Bureau of Transportation Statistics, *Air Carrier Summary Data (Form 41 and 298C Summary Data), T2: U.S. Air Carrier Traffic and Capacity Statistics by Aircraft Type, Aircraft Type Analysis,* Bureau of Transportation Statistics, Washington, D.C., March 2009, available at http://www.transtats.bts.gov/Fields.asp?Table_ID=254, accessed July 22, 2009.

NAOMS study team should have investigated the potential impact of these biases and evaluated alternatives such as the use of less stringent filters.

In summary, the NAOMS team faced substantial challenges in developing and implementing sampling designs for the AC and GA surveys. As with most real applications, the team had to make compromises in the final design—most notably in the development of the sampling frames. Stratification based on certification status and, to a lesser extent, the use of the public version of the ACD both introduced the potential for bias in results from the AC sample. While neither decision was made without reason, the NAOMS study team should have investigated the potential impact and magnitude of these biases in order to evaluate these and alternative decisions. Such analyses are critical both for understanding the value of the data being collected at that time and, more importantly, for learning how to improve the sample design for ongoing data collection.

4.3.3 Failure to Locate and Other Nonresponse Issues

Two steps during the field implementation of the two surveys may have contributed to additional coverage errors in the AC and GA samples. One was the failure to locate sampled pilots for whom telephone numbers could not be obtained. The second was noncompletion (refusal to participate) by pilots.

Because the questionnaires were administered by telephone, the project team needed current telephone numbers to contact sampled pilots. Telephone numbers were not available on the public ACD, so NAOMS used a service called Telematch to find telephone numbers based on names and addresses from the ACD, supplemented by change-of-address information from the Post Office. NAOMS tried to interview only those pilots for whom it could obtain telephone numbers, either from Telematch or in response to a mailing to the best address on record. These procedures yielded location rates of 82 percent and 71 percent, respectively, for the AC and GA samples. These rates are reasonable, given that the address information in the ACD was likely to be out-of-date.

Estimates of the rates of events will be biased if the pilots who were not located had rates of events substantially different from those of located pilots. The committee has no reason to expect that such a difference exists between located and nonlocated pilots, although additional data would be needed to verify that failure to locate is not a source of bias.

Nonresponse occurred at two other points for both surveys—at initial contact and after screening for eligibility. Because the NAOMS team could not know whether initial nonresponders were eligible, there was no way to compute response rates for eligible cases. Assuming independence of eligibility and initial response, the committee computed completion rates of 85 percent and 70 percent, respectively, for the AC and GA samples. This response rate for the AC survey is excellent by most standards for a survey of this length, possibly reflecting the interest level of commercial pilots in aviation safety. While lower, the GA response rate is also quite good. However, because the decision to cooperate with this type of survey might be influenced by recently experienced safety-related events, data comparing respondents with nonrespondents would be particularly valuable for assessing the potential bias due to nonresponse.

Taking into account both failure to locate and noncompletion, the estimated overall response rates were 69 percent[20] and 50 percent, respectively, for the AC and GA surveys.

Finding: The NAOMS team should have collected supplemental data to assess the potential biases related to opt-out issues, the certificate requirement, and nonresponse issues during the early phase of the survey. An analysis of the additional data would have provided a more reliable assessment of the various biases and may have led, if necessary, to the development of alternative sample design strategies to address the problems.

4.4 CROSS-SECTIONAL DESIGN VERSUS PANEL DESIGN

The term *cross-sectional design* refers to the selection of different samples of respondents at each time period. *Panel design,* on the other hand, involves the selection of groups of respondents who participate in the survey for a period of time (two or more successive periods). The NAOMS team evaluated both approaches during the first year of full operation of the survey by randomly assigning half of the participants to each design. The results of the first year indicated that the NAOMS survey would achieve a very high response rate and quality if the cross-sectional design was used.

The NAOMS team finally settled on a cross-sectional design. The committee agrees with this decision for several reasons:

- Response rates from panel designs are generally lower than those from similarly designed cross-sectional designs.[21] For example, respondents are less likely to participate in a survey if they are faced with the prospect of responding repeatedly to the survey over time. This problem may manifest itself as attrition, reducing follow-up response rates over time.
- A panel design would have risked "conditioning effects," by which a pilot's responses systematically change as a result of having taken the same survey previously. For example, respondents might become more sensitized to certain events after being asked about those events. Alternatively, a respondent who reports an event during one period may be less inclined to report a similar event in a later period because it seems "less interesting." While it is hard to know whether such effects would lead to more, or less, accurate reports, the possibility of their existence would complicate the interpretation of a survey with a primary goal of estimating trends.
- Finally, maintaining the confidentiality of pilots would be more difficult with a panel design.

Finding: The choice by the NAOMS team of a cross-sectional design over a panel design was appropriate.

4.5 RECALL PERIOD

The NAOMS project needed to specify an appropriate recall period (the previous *n* days from the date of the interview) for the survey. The NAOMS team conducted research as part of its planning process to determine some of the impacts of using a different recall period. As may be expected, the tests found that longer recall periods resulted in more reported events but lower rates of reported events, as well as a decline in the respondents' confidence in the accuracy of their responses. The team experimented with 30, 60, and 90 days in the initial stages of the study and eventually settled on a 60-day recall period.

---

[20] The committee's estimate of 69 percent differs from the results of the calculations conducted by the GAO because the latter calculation assumed (1) that all pilots who were located but not screened for eligibility were indeed eligible and (2) that the eligibility proportion was higher for nonlocated pilots than for those who were eventually screened.

[21] Nicole Watson and Mark Wooden, "Identifying Factors Affecting Longitudinal Survey Response," in Peter Lynn, ed., *Methodology of Longitudinal Surveys,* Wiley, New York, 2009.

The committee did not undertake a review of the methodology used by the NAOMS team to determine the recall period because it did not have access to data from these studies. Therefore, the committee cannot comment authoritatively on the choice of a 60-day recall period versus a recall period of a different length. However, analysis of the redacted survey data (discussed in Chapter 7) indicated several problems: (1) considerable rounding effects in reported numbers of hours and flight legs flown (see Figure 7.1 in Chapter 7) and (2) a high fraction of anomalous data for both Section A (number of hours/legs flown) and Section B (event counts) of the survey questionnaires. It is possible that these problems are not related to the recall period. Nonetheless, it is surprising that Battelle's final report notes the existence of a "downward bias" without a systematic investigation of the size of the bias, how it varies across the different types of events, and so on.[22] This information is critical to the validity of the survey results, especially given the rare nature of some of the events being surveyed. Finally, the committee agrees with the Battelle report that the effect of the bias would be smaller on trends than on actual rates, provided that the nature of the bias remains constant over time—hence the need to investigate the magnitude and nature of the potential biases.

4.6 DATA-COLLECTION METHOD

The NAOMS team considered three different ways to conduct the survey: in-person interviews, self-administered questionnaires, and computer-assisted telephone interviews. Each type was weighed against several criteria, including cost, respondent satisfaction, response rate, and quality of the data.

The NAOMS team tested the different methods of conducting the survey in a field trial. Early in the field trial, the NAOMS team determined that in-person interviewing required too much time and cost, and it was dropped from consideration. The results of the trial also demonstrated clear differences between the CATI system and the self-administered questionnaires. While the CATI system took longer to complete and cost more, it had a higher response rate and fewer incomplete responses. In the end, the team opted to use the CATI system for the full implementation of the survey.[23]

The NAOMS team decided to use professionally trained interviewers rather than aviation-safety professionals to conduct the interview. There are advantages and disadvantages to both options. The advantage of using aviation-safety professionals lies in their ability to clarify the intent of the questions or to ask the respondents to verify answers that seem implausible. However, there is the possibility of interviewer bias, by which the interviewer may lead the respondent in the direction of expected responses, and it would be difficult to quantify the nature of the assistance and clarifications provided by the interviewer and whether the interviews led to reproducible answers. For this reason, most surveys use professionally trained interviewers who have no subject-matter knowledge in the area of investigation. On balance, if the survey instrument is well defined and the possible questions and ambiguities are anticipated and addressed, the use of professionally trained interviewers will lead to more statistically reliable results.

Finding: The decision by the NAOMS team to use professionally trained interviewers was reasonable. The use of the CATI method for the survey was also appropriate.

---

[22] Battelle, *NAOMS Reference Report,* 2007, p. 27.

[23] Ibid., p. 31.

5

Analysis of NAOMS Questionnaires

The charge to the committee included an assessment of the design of the questionnaires used in the NAOMS survey. The structure of the two questionnaires used—one for AC pilots and one for GA pilots, as described in Chapter 4—was the same, with different questions as appropriate. Section 5.1 reviews the questionnaire structure, and Section 5.2 provides the committee's analysis of the questions. The complete AC and GA questionnaires can be found, respectively, in Appendixes G and H of this report.

5.1 QUESTIONNAIRE STRUCTURE

The NAOMS survey comprised four sections:

- *Section A—Flight Activity Levels and Background*—Section A included background questions on pilot and aircraft exposure information (number of legs and hours flown). Specifically, the data collected included number of flight hours, number of flight legs, aircraft size, propulsion type, flight type (domestic, international, etc.), crew role (captain, first officer, etc.), amount of pilot experience (in years), and mission type (passenger, cargo, etc.). The primary questions were the number of flight legs and flight hours flown by the respondent (pilot) during the recall period. This information provided the "exposure" variable (legs or hours) to be used as the denominators in the rate calculations.
- *Section B—Safety Related Events*—Section B asked the pilots about the number of events that occurred for a wide range of event types. These questions were designed to be asked routinely over a long period of time to enable the computation of safety event rates and event rate trends.

To select the topics in Section B, the NAOMS team first consulted "existing aviation safety data repositories maintained by NASA, the FAA, and the NTSB to identify known safety issues." The team also asked pilots and others in the aviation field about "safety issues important to them based on their first-hand operating experience."[1] This collection of information occurred through consultation with the ASRS analysts, AC pilot focus-group sessions, and two workshops hosted by the NAOMS team.

The focus-group sessions were held in the Washington, D.C., area in August and September of 1998. Battelle's final report states that the sessions included 37 active air carrier pilots flying both domestic and international routes,

[1] Battelle, *NAOMS Reference Report,* 2007, pp. 18-19.

but it does not indicate how these 37 were selected. Each session lasted 90 minutes, was led by a professional facilitator, and involved between 2 and 15 pilots. Pilots were encouraged to mention as many different types of events as possible, including any events that should not occur during normal operations. All sessions were recorded and later transcribed. The specific questions posed to the groups are in Appendix 6 of Battelle's final report.[2] The NAOMS team also conducted nine one-on-one interviews to identify additional events that did not surface in the focus groups.

The consolidated list of safety-related topics, as generated by the focus-group sessions and the interviews, is in Appendix 7 of Battelle's final report. Decisions on which topics to include in Section B of the NAOMS survey were based on "a desire to select events serious enough to be good indicators of the safety performance of the aviation system, yet not so serious that they would occur too rarely to be captured in the survey."[3] Some rare events were included in Section B because of strong industry interest in these specific topics.

The NAOMS team structured the organization of questions in Section B based on the team's research on how pilots organize their memories. The advice of "accomplished survey methodologists and aviation subject matter experts" was used "to craft questions responsive to each topic."[4]

- *Section C—Special Topics*—The questions in Section C on special focus topics were intended to be asked only over a few months or years and then replaced by new topics. Three different Section C question sets were developed for the AC questionnaire: one concerning minimum equipment lists, a second one addressing in-close approach changes, and the final one requested by a Commercial Aviation Safety Team subgroup of the Joint Implementation Measurement Data Analysis Team (CAST-JIMDAT) focusing on "the development of baseline aviation system performance measures."[5]
- *Section D—Questionnaire Feedback*—The questions in Section D provided respondents with a chance to give feedback regarding their survey experience to the interviewers who called. While not directly applicable for safety event rates or trends, some answers to these questions have provided possible topics for future surveys, should they occur.

During the field test of the survey, 9 percent of the pilots interviewed indicated that they found one or more of the questions confusing. Most of the confusion was related to Section C questions, particularly those on the minimum equipment list. For the full implementation of the survey, the project team decided not to use these questions.[6]

5.2 ANALYSIS OF THE QUESTIONS

On the basis of its assessment of the AC and GA questionnaires, the committee found four types of problems that reduced the usefulness of the data collected in the NAOMS survey:

1. The questionnaires were designed so that events and experiences from markedly different segments of the aviation industry were aggregated together (and cannot be disaggregated).
2. Some of the questions asked pilots for information that they would likely not have had without a post-flight analysis.
3. Some of the questions had vague or ambiguous definitions of what constituted an event to be measured.
4. Some of the questions did not have a clear link between the measured event and aviation safety.

These problems are discussed in detail in the following subsections. (While the examples shown below come primarily from the AC questionnaire, the general problems discussed exist in both the AC and GA questionnaires, unless otherwise specified.)

[2] Ibid., p. 19.
[3] Ibid.
[4] Ibid., p. 20.
[5] Ibid.
[6] Ibid.

5.2.1 Aggregation of Markedly Different Segments of Aviation

As discussed in Chapter 4, there are two issues with the sampling frame. One is whether the appropriate pilots were sampled; the other is whether the flight legs for which the pilots provided information were confined to those in the operations of interest. For example, in the AC survey, once pilots who were qualified to conduct Part 121 operations were selected, the survey should have restricted the flight legs, flight hours, and events reported to Part 121 operations and excluded flights and events that occurred in other operations. Unfortunately, this was not the case, as the survey did not ask for information about whether the pilots had actually flown in Part 121 operations, or whether they had flown in any other operations besides Part 121 during the recall period.

The AC field test survey did include a question (A3) asking the pilots how many hours and flight legs they flew in "Scheduled Major or National," "Scheduled Regional," "Unscheduled," and "Cargo" operations,[7] as well as the make and model of the aircraft that they flew in each of these four categories. However, the field test questions asked the pilot to report only those safety-related events that occurred in "commercial aircraft," so the questionnaire failed to link the safety-related event to one of the four categories of service or to the make or model of aircraft that was being flown when the event occurred. The final AC questionnaire did not include any question or reference to the four types of operations contained in the field survey or to other types of operations in which the pilots might have flown. Rather, the only reference was to the hours and legs flown as a "crewmember on a commercial aircraft" and to the aircraft makes and models that the pilot "flew commercially" during the recall period.

Several problems emerge from these aspects of the questionnaire. Even the term *air carrier* is very broad and includes not only the well-known scheduled passenger airlines such as American, Delta, and Southwest, but also the large air cargo airlines such as FedEx and UPS. These operations were lumped together in the NAOMS survey, but the distinction between passenger and cargo airlines is potentially important. As discussed in Chapter 2, the principal concern with aviation safety has been that of reducing fatalities, so the crash of a cargo plane has less potential for loss of life than that of a passenger plane. Also, these two segments of aviation may fly similar aircraft, but they fly in different operating environments and thus may well experience different rates of some of the safety-related events included in the NAOMS survey. Yet, the survey provided no way of distinguishing which operations or which safety-related events occurred in passenger versus cargo operations.

The term *air carrier,* as used in Federal Aviation Regulations (FAR; 14 CFR 1.1), also includes nonscheduled air taxi operators providing service in some very small aircraft in some very challenging operating environments (such as Alaska) as well as charter-flight operators and small, nonscheduled cargo aircraft operators. Many of these operations are not conducted under Part 121 rules and procedures, but rather under other, typically less strict, regulations, often under Part 135 and in very different operating environments. Thus the term *air carrier* includes an extraordinarily heterogeneous collection of operators, aircraft, and operating environments (see Appendix D for more information on the relevant parts of FAR).

The survey's use of the term *commercial* is even more problematic. In aviation, *commercial* is a broad term that could be interpreted to include an even wider variety of operations than would be included under the term *air carrier*. In addition to operations and aircraft flown under Part 121, commercial operations also include operations and aircraft flown under both Part 135 and even some operations, such as corporate aircraft, under Part 91.[8] It is possible that the pilots may have interpreted commercial operations to include Part 136 (commercial air tours), Part 137 (agricultural operations), Part 141 (pilot schools), and perhaps other segments of aviation as well. The aircraft typically flown under Parts 135 and 91, as well as under Parts 136, 137, and 141, are smaller and often not as well equipped as those flown under Part 121. Moreover, the types of flights in these operations and the environments in which they occur are usually quite different from Part 121 air carrier operations (see Appendix D for more information on the relevant parts of FAR).

Based on an analysis of the available survey data, the committee determined that about 75 percent of the pilots in the AC survey reported flying only one make and model of aircraft during the recall period. As might

[7] Joan Cwi, Director, Survey Operations, Battelle Memorial Institute, *The NAOMS Field Trial,* presentation to the NAOMS Working Group Meeting, Battelle Memorial Institute, Columbus, Ohio, December 18, 2003.

[8] *Code of Federal Regulations, Title 14: Aeronautics and Space*, December 2005, available at http://ecfr.gpoaccess.gov/cgi/t/text/text-idx?c=ecfr&rgn=div8&view=text&node=14:1.0.1.1.1.0.1.1&idno=14, accessed July 20, 2009.

be expected, given the sampling frame for the pilots, the most frequently reported types of aircraft were those in use by the major scheduled passenger airlines—such as B737s, B747s, B777s, B757s, and A320s. Some pilots reported flying DC-10s, MD-11s, B727s, and DC-8s, aircraft that are not used in scheduled passenger service but that are or were at the time used frequently in scheduled cargo service by FedEx and UPS.

However, several pilots in the AC survey reported flying all of their time in aircraft such as B707s, Gulfstream, and Learjet aircraft, yet these aircraft are not reported in the Bureau of Transportation Statistics data to have been used by air carriers operating under Part 121 during the period covered by the survey.[9] For example, both Gulfstream and Learjet are used extensively as corporate aircraft flown under Part 91. The inclusion of these aircraft in the survey appears to indicate that pilots interpreted the term *commercial* more broadly than the FAA definition of the term *air carrier* and certainly to include more than Part 121 operations. Since these pilots who flew only one of these aircraft during the recall period (B707, Gulfstream, and Learjet) assumed this broad interpretation of *commercial,* it is likely that other pilots who flew more than one aircraft type during the recall period used a similarly broad definition of *commercial* in their responses.

There are three troubling aspects about the survey's failure to distinguish between safety-related events occurring while flying different aircraft, under different regulatory regimes, and in different operating environments:

1. The accident rates have varied considerably across these different segments of aviation, so one might expect the frequency of safety-related events also to vary across these segments. For example, based on NTSB accident data for 1989 through 2008, the rate of fatal accidents per 100,000 flight hours was more than 33 times greater for Part 135 operations than for Part 121 scheduled airline operations. During the same period, the fatal accident rate for general aviation was more than 91 times greater than for the scheduled airlines.[10] It is possible that similar differences also exist in rates of safety-related events.

2. The growth rates of these industry segments have been different, so over time, the mix of pilots in the sample who operate in these different segments is likely to change. For example, during the 1989 through 2008 period, Part 121 scheduled airline flight hours increased 77 percent, while Part 135 flight hours decreased 25 percent and general aviation[11] flight hours declined 21 percent.[12] Thus the inability to link the safety-related event either to the aircraft type or to the type of operating environment would seem to severely hinder, or more likely prevent, any meaningful analysis of event rates by aircraft type or type of operation. Moreover, because the mix of operations included in the NAOMS aggregate rates is likely to change over time, trends in the NAOMS aggregate rates would not necessarily reflect trends in the occurrence of these events in the airspace system. Instead, they might reflect a change in the mix of pilots flying in markedly different operating environments.

3. Finally, these limitations severely hinder or more likely prevent any meaningful comparison of event rates or trends in those rates calculated from the NAOMS data with event rates derived from other sources of data such as those compiled by the FAA.

The same basic problem was found in the GA questionnaire. The GA questionnaire did ask pilots what proportion of their flight hours and flight legs were done under Part 121, Part 135, and Part 91 and what aircraft types they flew in each of these segments. Then, the questionnaire asked respondents to report only safety-related events for those flights that occurred under either Part 135 or Part 91. But, as in the AC survey, the GA survey failed to link the events either to the type of operation or to the type of aircraft, so the GA survey aggregated events and

---

[9] Bureau of Transportation Statistics, *Database: Air Carrier Statistics (Form 41 Traffic)—U.S. Carriers,* Research and Innovative Technology Administration (RITA), available at http://www.transtats.bts.gov/Tables.asp?DB_ID=110&DB_Name=Air%20Carrier%20Statistics%20%28Form%2041%20Traffic%29-%20%20U.S.%20Carriers&DB_Short_Name=Air%20Carriers, accessed June 11, 2009.

[10] These fatality rates were from committee calculations using data found in NTSB, *Aviation Accident Statistics*, NTSB, Washington, D.C., 2009,Tables 6, 8, 9, and 10, available at http://www.ntsb.gov/aviation/stats.htm, accessed July 15, 2009.

[11] The Department of Transportation considers general aviation to be operations of U.S.-registered civil aircraft not operated under FAR Part 121 or Part 135.

[12] These growth rates were from committee calculations using data found in NTSB, *Aviation Accident Statistics,* 2009, Tables 6, 8, 9, and 10.

legs from both Part 91 and Part 135, and these have been seen to be very different operating environments. Thus, the same three concerns that were raised for the AC questionnaire are also relevant for the GA questionnaire.

Finding: Both the air carrier and the general aviation questionnaires asked respondents to include events, flight hours, and flight legs in segments of aviation that went beyond even the broadest definition of AC operations and beyond the conventional definition of GA. As a result, highly disparate segments of the aviation industry were aggregated into the safety-related event rates that were calculated from these surveys.

Finding: The inability to link the safety-related event to the aircraft type or to the operating environment in which the event occurred severely hinders any meaningful analysis of event rates or trends in event rates by aircraft type or by segment of aviation.

5.2.2 Asking Pilots for Information That They Would Not Have Had Without Post-Flight Analysis

Some of the questions in both the AC and GA questionnaires asked the pilots about causes of events. As will be discussed below, the pilot might well perceive that an event occurred or that there was a specific cause for the event based on information available in the cockpit at the time. However, in many situations, only a post-flight analysis of the flight data recorder or of the aircraft itself would reveal what the event actually was or what had caused the event. In air carrier operations, pilots would not typically have access to that information. Thus, many pilots would be responding to the survey on the basis of their *perception* of what had occurred rather than on the basis of what the post-flight analysis showed to actually have occurred. This is particularly problematic if the data from these types of questions are then compared as actual events with aviation safety data from other sources.

For example, the pilot might see indications in the cockpit consistent with an engine failure and thus perceive that there had in fact been an engine failure, but the actual failure could have been something else. An accessory or component failure that reduced thrust or revolutions per minute could cause the pilot to perceive an engine failure, whereas analysis might show that the engine itself did not fail. Similarly, a Full Authority Digital Engine Control failure could shut down the engine, resulting in what appeared to be an engine failure to the pilot. A post-flight analysis of the flight data recorder or the aircraft would reveal what had actually happened, but few pilots would have access to this information. The pilots would therefore be answering the question with a broader definition of *engine failure* that would include more kinds of events than the definition used by the FAA and by other sources of data. Comparing the data as if the terms were the same would be misleading and would almost certainly result in NAOMS reporting higher rates of *engine failure* than the other sources would report. Such comparisons are inevitable. Indeed, the NAOMS team's preliminary analysis of the data and the presentation based on that analysis made this comparison[13] and drew the inference, later reported by the media,[14] that the FAA was under-reporting this event. If the NAOMS survey was going to use the same terms as those used by other established data sources, the committee believes that it should have used the same definitions. Otherwise, it should have recognized the difference in definitions by using different terminology.

Pilot perceptions of safety-related events can provide valuable information. Indeed, some surveys are designed to collect data on the respondents' perceptions or opinions of events. However, NAOMS was not conceived or justified on that basis. Rather, its stated intent was to provide information about the rates of specific events that are related to safety. The NAOMS survey was justified in the expectation that it would be a new tool that had been missing within the aviation safety field—a tool that could generate statistically valid rates of events and track trends over time for the entire NAS. NASA's intent was to offer policy makers statistically valid estimates that would address the performance and safety of the entire NAS and would measure the impacts of various new policies and programs.[15]

---

[13] "Comparison Charts," presentation to the NAOMS Working Group Meeting, Washington, D.C., May 5, 2004, p. 2.

[14] Alexandra Marks, "NASA plays down its air safety report," *Christian Science Monitor,* January 3, 2008.

[15] Irving C. Statler, ed., *The Aviation System Monitoring and Modeling (ASMM) Project: A Documentation of Its History and Accomplishments: 1999-2005,* NASA, Washington, D.C., June 2007, available at http://www.nasa.gov/pdf/225024main_TP-2007-214556%20ASMM_Project.pdf, accessed June 11, 2008, p. 17.

Following are examples of NAOMS survey questions with the problem discussed above. Question ER5 in the AC questionnaire reads:

How many times during the last (TIME PERIOD) did an inflight aircraft on which you were a crewmember experience smoke, fire, or fumes that originated in any of the following areas:

A. the engine or nacelle?
B. the flight deck?
C. the cargo hold?
D. the galley?
E. elsewhere in the passenger compartment?
F. During the last (TIME PERIOD), how many times did an inflight aircraft on which you were a crewmember experience smoke, fire or fumes that originated other than in the engine or nacelle, flight deck, cargo hold, galley, or passenger compartment?

1. Where did the smoke, fire or fumes originate? SPECIFY.

This question cannot be answered accurately without analyzing post-flight data because pilots may or may not know or be able to tell the difference between bleed air fumes (oil-based), electrical fumes, or solid object fumes. For example, smoke or fumes detected by the pilot anywhere in the aircraft could have originated in the engine and spread throughout the aircraft as a function of pressurization bleed air extraction. In some situations, it is also possible that the perceived smoke or fumes could have come from outside the aircraft, particularly when the aircraft is on a taxiway adjacent to an active runway awaiting its turn to take off.

Other questions have similar potential problems. One question asks about uncommanded movements of control surfaces, but the pilot would not necessarily know what failure resulted in what appeared to be an uncommanded movement. Another question asks for how many degrees an aircraft rolled in a wake turbulence encounter, but without a post-flight analysis of the flight data recorder, a pilot would not know how much the aircraft had rolled. Another question asks for airspeed deviation during a wind shear event. During such sudden and unexpected encounters, pilots are typically more concerned with recovering the aircraft than with estimating the degree of roll or airspeed deviation. Other questions ask whether the aircraft came within 500 feet of another aircraft. Again, in such unexpected situations, the pilot is typically neither in a position nor trained to make an accurate estimate of the absolute distance. Still another question asks whether hazardous materials were packaged and loaded on the aircraft in compliance with the appropriate regulations, but those are not regulations that a pilot is required or even expected to be familiar with. Appendix F contains nine additional examples of questions that potentially ask pilots for information that they would not typically have access to.

Finding: Some of the questions in the NAOMS survey would not provide accurate and consistent measures of events because they asked about situations in which pilots would not typically have access to the information needed for an accurate response.

5.2.3 Problems with Structure and Wording of the Questions

The literature on the design of survey questionnaires emphasizes the importance of clear and carefully worded questions to elicit reliable responses. In particular, it is important that respondents answer a question consistently even if the question is asked at different points in the survey and that, ideally, different respondents interpret the same question in the same way. This is especially important in a survey such as NAOMS that is intended to collect information on events that occurred, rather than to collect opinions and perceptions of respondents.

The NAOMS team field-tested the survey at several stages and apparently redesigned the questions to take into account some of the comments that it received. Nevertheless, the committee finds that several questions in the survey contain wording that pilots may have found difficult to interpret precisely and to answer consistently. These include (1) long questions with complex structure that would be difficult to understand in a computer-assisted telephone interview; (2) questions that appear to combine multiple, unrelated events; (3) questions about events that are not well defined; and (4) questions containing vague terms. As is discussed below, these problems may have resulted in inconsistent responses and led to additional measurement error in some of the responses.

5.2.3.1 Complex Structure

The following is an example from the NAOMS survey of a long question with complex structure:

> AC2. During the last (TIME PERIOD), how many times did an aircraft on which you were a crewmember perform an evasive action to avoid an imminent in-flight collision with another aircraft that was never closer than 500 feet including evasive action in response to a TCAS advisory?

This question includes several conditions—time period, evasive action, imminent collision, 500 feet—that the respondent must keep in mind while deciding on an answer. Doing so is particularly difficult in a telephone interview. It is not easy to digest such questions even if the interviewer repeats the question, which would only be done at the respondent's request. The literature on questionnaire design clearly recommends against using such questions.[16]

5.2.3.2 Combining Multiple Events or Causes

Several of the NAOMS survey questions had two or more subparts, and it would be unclear to the respondent which part to answer. This is sometimes referred to in the literature as a *double-barreled question*.[17] Two examples are given below.

> AT2. How many times during the last (TIME PERIOD) did an aircraft on which you were a crewmember fly at an undesirably high altitude or airspeed on approach due to an A.T.C. [Air Traffic Control] clearance?

This question combines two events: undesirably high altitude and undesirably high airspeed. It is unclear why one should be interested in the total number of the two events, as their causes and consequences are likely to be different. Such questions also create a problem for data analysis, as an answer of *X* in this example can mean *X* times high altitude, *X* times high airspeed, or some combination of those possibilities.

> AH2. During the last (TIME PERIOD), how many times did an aircraft on which you were a crewmember accept an A.T.C. clearance that the aircraft could not comply with because of its performance limits?

This is another example of a multipart question. The answer can refer to two very different situations from an aviation safety perspective. In the first case, a pilot may have accepted the clearance and subsequently determined the inability to comply. In the second, the pilot may have accepted the clearance knowing in advance that the aircraft could not, or reasonably might not, be able to comply with the clearance. From the standpoint of trying to reduce such potentially unsafe events, it is critical to distinguish between these two different situations.

5.2.3.3 Unclear Definition of Events

Some questions in the NAOMS survey asked the respondents about events that were not clearly defined. For example:

> AH9. During the last (TIME PERIOD), how many times did an aircraft on which you were a crewmember experience a hard landing?

There was no definition of what constitutes a *hard landing* and there was no consensus on the meaning of the term even among the pilots within the committee.

Several questions included the phrase *near collision* or *nearly collide*. The term *near collision* is difficult to quantify, and there is likely to be considerable variation among the respondents in interpreting it. It may have been

---

16 Bradburn et al., *Asking Questions,* 2004.

17 Ibid.

better to ask for the number of times that a near collision had led to some specific action on the part of the pilot, such as evasive action or the reporting of the event.

5.2.3.4 Use of Vague Terms

There are several questions in the NAOMS survey with vague modifiers, such as *abrupt, accurate, severe,* and *time-critical.* Some examples are given below.

> TU1. During the last (TIME PERIOD), how many times did an aircraft on which you were a crewmember encounter severe turbulence that caused large abrupt changes in altitude, airspeed, or attitude?

Different pilots would interpret the phrases *severe turbulence* and *large abrupt changes in altitude, airspeed, or attitude* differently. *Severe turbulence* has a formal definition in the FAA's *Aeronautical Information Manual:* "Turbulence that causes large, abrupt changes in altitude and/or attitude. It usually causes large variations in indicated airspeed. Aircraft may be momentarily out of control."[18] However, this definition also uses vague modifiers, which would prevent consistency in the answers from the respondents. Moreover, *severe* is the third of four levels of turbulence defined in the manual—light, moderate, severe, and extreme—all of which use vague modifiers in their definitions. Survey respondents familiar with these four terms and definitions might well have found it ambiguous as to whether events that they perceived to fit the definition of extreme turbulence were to be included in their response to the preceding question.

> WE1. During the last (TIME PERIOD), how many times did an aircraft on which you were a crewmember lack accurate weather information when crewmembers needed it while airborne?

The perception of the extent to which the weather information was accurate or sufficiently accurate for their needs is likely to vary among respondents.

The problematic types of questions exemplified above must be carefully examined using cognitive testing techniques,[19] and the committee did not see any evidence that this had been done. It is possible that some of these questions could not be worded more precisely, in which case they should not have been included in the survey.

Appendix F provides additional examples of questions identified by the committee as having one or more of the deficiencies discussed in this section.

Finding: There are several problems with the structure and the wording of the survey questions. These problems may have led to varying interpretations and judgments, thus reducing the value of some of the survey results.

5.2.4 Questions About Events Without a Clear Link to Aviation Safety

Finally, in reviewing the questionnaire, the committee was concerned about the lack of relevance to aviation safety of the series of questions about in-close approach changes in Section C of the AC questionnaire. The question on which Section C is based is as follows:

> IC1. During the last (TIME PERIOD), how many times did an aircraft on which you were a crewmember receive an unrequested clearance change to runway assignment, altitude restrictions or airspeed within 10 miles of the runway threshold?

The committee questioned the relevance of this section to aviation safety because of the 10-mile criterion that was established as the basis for the question. A 10-mile criterion for defining something as "safety related" seems ad hoc and inconsistent with other definitions of safety around the terminal area. By including these questions as

---

[18] Federal Aviation Administration, *Aeronautical Information Manual,* Washington, D.C, February 14, 2008, Section 7.1.23 PIREPS Relating to Turbulence, Section 7.1.25 Clear Air Turbulence (CAT) PIREPS, and Table 7-1-9 Turbulence Reporting Criteria Table.

[19] Robert M. Groves, Floyd J. Fowler, Jr., Mick P. Couper, James M. Lepkowski, Eleanor Singer, and Roger Tourangeau, *Survey Methodology,* Wiley, New York, 2004, p. 213.

part of "safety-related events," some respondents might deduce that there is something inherently unsafe about an approach change inside of 10 miles if the crew did not request it. The response to this question and whether or not there was a potential safety concern could vary greatly, depending on where the change was initiated and on how much of a heading change would be required in the maneuver. Similarly, the question did not allow for the consideration of the type of aircraft involved, since smaller, narrow-body aircraft could more easily accept a change much closer to the runway than could a wide-body aircraft.

6

The Redacted Data and Their Limitations

Task 4 of the charge to the committee asks for "an analysis of the NAOMS project survey data provided by NASA to determine its potential utility." This chapter focuses on the limitations of the redacted data that have been released to the public.

To maintain the confidentiality of the survey participants, NASA released only redacted versions of the survey data. Two versions of these redacted data are currently available: Phase 1 was released on December 31, 2007 (with an update, Phase 1a, on February 6, 2008); Phase 2 was released on September 30, 2008. NASA has not officially released any analyses based on the NAOMS survey data.

Only data from the air carrier (AC) survey are discussed here, because the survey of general aviation (GA) pilots lasted for only 9 months. Nevertheless, a number of the conclusions made here, such as those on the limitations of the redacted data and data quality, will also apply to the survey results from GA pilots.

6.1 PHASE 1 AND 1a REDACTIONS

NASA characterized the initial release of data in Phase 1 as "conservative to ensure the responses do not contain confidential commercial information or information that could compromise the anonymity of individual pilots."[1] The strategies for redaction included the reordering, generalization, disaggregation, deletion, and/or editing of the survey responses. In particular, the recall period and recall date were removed, and the legs flown were disaggregated from the majority of responses. The structure of the released data is discussed below.

Because Phase 1a was a relatively minor update of the release in Phase 1, it is not discussed here. (The main point of this update was the reclassification of 407 AC survey responses that had been filed with the GA responses. These respondents were contacted during the screening for the GA survey, but they actually were air carrier pilots and so were given the AC questionnaire instead of the GA questionnaire. Phase 1a also included an additional 701 GA rotorcraft responses.)

A detailed description of Phase 1 data and the modifications follows:

[1] NASA, National Aviation Operations Monitoring Service (NAOMS) Information Release, available at http://www.nasa.gov/news/reports/NAOMS.html, accessed April 14, 2008; see section headed "December 31, 2007."

- *Section A—Pilot background questions and relevant exposure information.* The original responses from each pilot are categorized to the following groups:

1. Time of Interview by Year (4 levels): 2001, 2002, 2003, 2004.
2. Flight Hours (5 levels): Less than 51, 51-90, 91-130, 131-170, Greater than 170 (hours)
3. Aircraft size (4 levels): Small or Other, Medium, Large, and Wide-body with only the aircraft that was flown the most being reported
4. Propulsion (2 levels): Turbofan and Turboprop or Other
5. Flight Type (3 levels): Domestic, International, and Unknown
6. Crew Role (2 levels): Captain and First Officer or Other
7. Amount of Pilot Experience (3 levels): Low, Medium, and High
8. Mission type (2 levels): Passenger and Cargo or Other

Phase 1 provides pilot responses on flight hours (item 2) but not on flight legs (see Section 6.2).

- *Section B—Counts for safety events.* The responses are provided in their original raw form except for the following modifications:

1. Responses that were considered to be rare (occurred in less than 0.1 percent of the surveys) were removed and were given in separate tables without linking them to the pilot response for other events.
2. "High unique" response values were replaced with the next closest numerical value in that field. "High unique response" denotes cases where a pilot's response for an event count was unusually high in comparison to the responses of the other survey participants.

- *Section C—Responses to special questions concerning baseline performance measures and "in-close" changes to approach and landing.* High unique and rare counts are redacted in the same manner as in Section B. Free-text responses are aggregated into a separate file.
- *Section D—Pilot feedback on the questionnaire.* The numerical values are reported with free-text responses again aggregated into a separate file.

In addition, files of partial raw responses for aircraft type flown, complete set of hours and legs, and career hours flown are provided separately.

Other features of Phase 1 redaction include the following:

- Individual pilot responses in Sections A and B are linked by means of a uniquely assigned *Random Identification Number*. This permits the examination of all the responses from a particular pilot in the analysis.
- The main release was cleaned, with those rows having missing data or outliers reported in separate files.[2]

It appears that the redaction strategies were developed primarily with confidentiality issues in mind. The following comments discuss the advantages and disadvantages from the viewpoint of information value in the redacted data.

---

[2] "Prior to the recent redaction steps taken, NAOMS air carrier survey responses were evaluated by Battelle at two stages. During initial processing, Battelle refined the set of survey responses using a technique called the Chebyshev process and related criteria to remove 322 responses of doubtful quality to avoid contaminating analyses of the responses. Battelle cites the following specific reasons for their removal: 1) number of flight hours too small; 2) unreasonable ratio of hours-to-legs; 3) unreasonable responses to multiple questions; and/or 4) Section B not completed. These responses are provided below and are identified as the 'Outlier Survey Responses.' A second refinement was then made by Battelle during subsequent tabulation activities when the NAOMS project team sought a set of responses with all explanatory flight activity variables present (no null values in Section A fields). This was done to ensure that all tabulations on the responses based on flight activity fields had a consistent total. This resulted in an additional 335 responses being removed. These responses are provided below, identified as the 'Survey Responses with Unknowns in Flight Activity Fields.'" NASA, *NAOMS Air Carrier Survey Responses*, available at http://erc.ivv.nasa.gov/news/reports/NAOMS_air_carrier_survey_data.html, accessed June 11, 2008.

There are several good features in the Phase 1 redacted data:

- The ability to link responses across sections so that all the numerical responses can be traced to a particular (anonymous) pilot during analysis;
- Post-processing of data to remove outliers before release (although the committee cannot comment on the validity of the method); and
- A separate file with actual (not categorized) flight hours, flight legs, and primary aircraft type for all respondents.

However, there are also several deficiencies in the Phase 1 redacted data:

- The grouping of the time of response into years is too coarse to permit sensitive analyses of trends over time.
- Grouping the data on "number of hours flown" reduces the ability to calculate event rates by the different explanatory variables, such as type of aircraft, pilot experience, and so on.
- The lack of information for number of flight legs flown is a serious limitation of this release. This is the right exposure variable for several event types, and one cannot compute event rates without this information.
- There is no way to judge the effects of the modification that was used to adjust for rare or high unique events.
- No information was provided on whether the NAOMS team made any attempts to identify *causes* for "outliers" that were removed. Removing outliers without assignable causes can lead to biased analyses that depend on the thresholds used to identify the outliers.

These issues and their consequences are discussed in more detail in Section 6.3.

6.2 PHASE 2 REDACTION

As of the release of this report, Phase 2 is NASA's last release of redacted NAOMS survey data and documents. The intent in this version was to release "the maximum amount of survey information"[3] without compromising (or at least minimizing the threat of compromising) the anonymity of the pilots or releasing confidential commercial information. Data on all 26,168 records were released. This release took a very different form from that of Phase 1; rather than relatively few spreadsheets, this release included more than 100 separate files. The tables included responses that were deleted or separated from Phase 1, such as incomplete survey responses, rare events, and high unique events.

The structure of the Phase 2 data is as follows. For each event in Section B of both questionnaires (for example, AC1 = number of bird strikes), a file with the following information is provided:

- Column 1: ID number: 1, 2, . . . , 26,168 (in random order that varies across the files for different events and hence cannot be linked).
- Column 2: Time of survey grouped into 4 years: 2001, 2002, 2003, 2004.
- Column 3: Number of hours flown in recall period, grouped into six categories: less than 46, 46-70, 71-100, 101-120, 121-150, greater than 150 (this is a different grouping from the one used in Phase 1).
- Column 4: Number of flight legs flown in recall period, grouped into five categories: less than 14, 14-22, 23-36, 37-60, greater than 60 (this information was not provided in Phase 1).
- Column 5: Aircraft type, grouped into 34 categories (more categories than in Phase 1).
- Column 6: Number of events (say, bird strikes) reported by the pilot as having occurred during recall period (data were not modified for rare and unique events as in Phase 1).

[3] Ron Colantonio, NASA Glenn Research Center, "National Aviation Operations Monitoring Service (NAOMS) 2008 Information Release Project," presentation to the NRC Committee on NASA's NAOMS Project, October 13, 2008, slide 7.

A separate file with the actual number of hours and number of flight legs flown by year for each respondent is provided, so the joint distribution of these two responses is available.

The advantages of Phase 2 redaction over that of Phase 1, from the viewpoint of information value, include the following:

- No redaction of Section B events;
- Availability of data for both exposure variables—hours (six categories) and flight legs (five categories)—for each safety event, as well as primary aircraft type; and
- Grouping of aircraft type into finer levels (34 categories).

However, there are also many problems, some of which are in common with Phase 1:

- The grouping time of the survey was in years rather than on a finer scale; the problem with this coarse grouping was discussed previously for the Phase 1 data.
- Exposure (in flight legs and hours) is coarsely categorized; the categories for hours do not correspond to those for Phase 1, which was apparently deliberate.
- Information for each safety event is given in separate files that cannot be linked. This was done to address the privacy/confidentiality concerns, but it does not allow the user to link information across multiple safety events—for example, if a single pilot reported multiple event types. This information would be especially useful in detecting aberrant data points.
- Some of the aircraft do not fit clearly into one of the 34 categories. (Even the pilots on the committee could not map all of the aircraft into the 34 categories consistently.) For example, B707 is not a category, but the aircraft does not really fit into any of the existing 34 categories.
- Respondents who "refused to answer" or "did not know" and fields with "missing information" are coded (999, 998, 997) or (99, 98, 97) or (9, 8, 7). But 9, 8, and 7 could be possible values for the responses, so a "missing code" of 9, 8, or 7 cannot be distinguished from a response of "9," or "8," or "7," reported events. The survey questionnaires include the codes used for each question,[4] but cross-referencing each response from the data set with each question in the survey is impractical.

6.3 LIMITATIONS OF THE REDACTED DATA

6.3.1 Effect of Grouping Time of Survey into Years

The categorizing of the time of the survey response by years in the redacted data of the NAOMS survey is too coarse to be very useful. The problem is exacerbated by the fact that pilots are recalling events that occurred within a recall period ranging from 30 to 90 days. So a non-negligible proportion (about 1/6 for a 60-day recall period) of the events reported in one year (say 2003) could have occurred in the previous year (2002). Such a level of ambiguity with respect to the time of the events makes it difficult to analyze the redacted data to achieve the objectives of NAOMS, which include (1) determining the effect of changes to airline safety procedures from the data and (2) tracking changes over time. For example, an event that had a seasonal trend (such as more bird strikes in the summer) cannot be captured from yearly data.

Obviously, there is a trade-off in retaining information for data analysis versus maintaining the confidentiality of pilots, and there has been considerable work in this area.[5] Since the committee does not have access to the original data, it cannot comment on the appropriate degree of categorization to achieve the best trade-off. From a

[4] NASA, *National Aviation Operations Monitoring Service (NAOMS) Phase 2 Information Release Survey Response Disclaimer,* Washington, D.C., September 30, 2008, p. 2, available at http://www.nasa.gov/pdf/279939main_Phase%202%20Release%20Summary%20092408_Final_2(508).pdf, accessed July 20, 2009.

[5] See American Statistical Association, *Privacy and Confidentiality*, Alexandria, Va., 2003, and references therein, available at http://www.amstat.org/committees/cmtepc/index.cfm?fuseaction=main, accessed July 15, 2009.

data analysis perspective, however, grouping by years is much too coarse and substantially undermines the goal of providing public data for independent analysis.

6.3.2 Effect of Grouping the Number of Hours and Number of Flight Legs Flown

The number of hours flown and the number of flight legs flown are the primary measures of exposure for calculating event rates. Depending on the type of safety event, one or the other is the right denominator for calculating event rates. In the absence of additional information on the categorization, one would typically use a surrogate value (such as the midpoint of the interval) as proxies for all of the values in that interval. However, the redacted data provided raw data on the numbers of hours and flight legs flown in a separate file. This allows one to compute the total numbers of hours and flight legs flown and hence the rates for each category exactly—calculating the total number of events in a category and dividing by the total number of hours or flight legs flown in that category to get rates for each category.

However, it is not possible to compute the total numbers of hours or flight legs flown by various subpopulations of interest from the redacted data. Consider the two subpopulations of AC pilots corresponding to domestic fights and to international flights. The redacted data in Phase 2 do not provide the total numbers of hours and flight legs flown for these two subpopulations, so one cannot compute the event rates. Various types of approximations are possible: for example, use a surrogate such as the midpoint of a category for the actual but unknown values in that category. However, all of these approximations will lead to additional uncertainty in the estimates. There are other subpopulations, such as the experience level of pilots and aircraft type, for which event rates would also be of interest but cannot be computed from the redacted data.

Phase 2 provides the total numbers of hours and flight legs flown for one particular subpopulation—aircraft make and model. Similar tables could easily have been provided for other, selected subpopulations without any danger of sacrificing respondents' confidentiality. It appears that the data redaction efforts did not fully take into consideration how the data would be analyzed and used.

6.3.3 Information on Size of Aircraft Flown

The redacted survey data provide the size of only one (primary) aircraft for each respondent: the aircraft that the pilot reported as having flown the most, grouped into four categories in Phase 1. This raises the possibility that the events reported may not be traced to the aircraft size. However, Table 6.1 shows that about 75 percent of the pilots reported flying 100 percent of the time in one aircraft. Even for the remaining 25 percent, it is likely that some fraction of them would have flown in the same size aircraft. Thus, the redaction does not appear to pose a serious problem for estimating event rates by aircraft size.

6.3.4 Grouping of Other Variables

Some other variables also exhibited the "most often" limitation. For example, a pilot is reported as flying "domestic" if that is what that pilot mainly flies. If most pilots fly either domestic or international but rarely both, the effect of this simplification (to just "domestic" or "international") would not be serious.

TABLE 6.1 Percentage of Time That Pilots Reported Flying in a Primary Aircraft and Corresponding Proportion of Respondents

Percentage of Time in Primary Aircraft	Proportion of Respondents
100%	0.743
90% to less than 100%	0.054
75% to less than 90%	0.070
50% to less than 75%	0.131
Less than 50%	0.002

6.3.5 Modification of Counts of Rare and High Unique Events in Phase 1

Many of the event types in the NAOMS survey are rare (that is, for many of the event types, pilots rarely reported the occurrence of the events). Modifications to the numbers of rare and high unique events in the redaction can severely alter the information in the data. There is no way to know if the reported non-zero counts referred to genuine events that should not have been removed, or if they are outliers that should be ignored (for example, typographical errors). Replacing the large values with the next closest numerical value leads to an underestimation of the event rates. It is difficult to assess the degree of this problem from the Phase 1 redacted data.

6.4 DATA ANOMALIES

The data values released in Phase 2 were not cleaned or modified for other unusual features, as they were in Phase 1. An examination of these raw data shows that an unusually high proportion of the numerical values are implausible, both for event counts (numerators) and number of legs/hours flown (denominators). This problem is discussed in more detail in the next chapter.

If the complete record for each respondent were available, it might be possible to detect whether a particular record stands out as an outlier. Specifically, if a respondent consistently reported implausible values for many questions, one might have some justification for removing the record. However, the complete set of responses from each survey is not available in Phase 2; rather it was split into separate files for each event.

Finding: The redacted data from the NAOMS survey have several limitations that further constrain the ability to analyze the data to meet the study objectives. The nature of the redaction differs in its two phases, so the type and severity of the limitations vary. The time of survey response is grouped into years (this feature is common to both phases), so estimates of event rates can be computed only by years. This limits the ability to (1) track the changes in event rates over shorter timescales, (2) determine the effects of changes in the aviation system on event rates, and (3) assess seasonal and similar types of effects. Grouping the exposure data (number of hours/legs flown) into categories increases the uncertainty of the estimates of event rates broken down by key characteristics, such as pilot experience. The separation of records into different files also constrains one's ability to detect anomalous records and thereby apply methods that could improve data quality.

The NAOMS data redaction efforts released by NASA resulted from a Freedom of Information Act (FOIA) request and the publicity that resulted from the denial of the request for the purposes of protecting the anonymity and confidentiality of the respondents (which was promised during the survey). The 1999 briefing to the ASRS Advisory Subcommittee asserted that "Participant Confidentiality is assured . . . . It will have *no means* of tracing a survey response to the individual who provided it; neither FOIA nor discovery actions will pose a confidentiality risk to NAOMS."[6] It appears that the NAOMS management team anticipated neither the need to eventually release the survey data to the public nor the consequent problems that would develop. (As some of the project team members noted in presentations to the committee[7] and as NASA noted in its response to the GAO report on NAOMS[8] a possible reason is that NAOMS was viewed, at least by some, as primarily a research study for developing a methodology. However, the project's submission for clearance from the Office of Management and Budget, dated June 12, 2000, clearly characterizes the project as a data-collection effort, not just a research study.[9])

It is clear that data analysis, reporting, and other post-survey activities were not adequately planned or properly anticipated. As a result, NASA appears to have rushed into developing after-the-fact redaction strategies for

[6] NASA, "NAOMS Development and Proof of Concept," presentation by NAOMS management team to ASRS Advisory Subcommittee, November 13, 1998, slide 15, available at http://www.nasa.gov/pdf/209893main_1998-11-13%20ASRS%20Advisory%20Subcommittee.pdf, accessed July 20, 2009.

[7] Connors, presentation to NRC Committee on NAOMS, 2009, p. 1.

[8] Government Accountability Office, *Aviation Safety,* 2009, p. 90.

[9] NASA, *Request to Conduct Federal Agency Survey, National Aviation Operations Monitoring Service (NAOMS)*, OMB Number 2700-0102, Washington, D.C., June 12, 2000.

releasing the data. The committee does not know the specific trade-offs that were considered and ultimately made in arriving at the redaction strategies. It can comment only on the severe negative impact of the chosen redaction strategy on the data analysis.

The issue of preserving the privacy and confidentiality of survey participants is not a new problem. It has been studied extensively, and considerable literature was published on the topic even prior to 2000.[10] In fact, many federal agencies have faced this problem regularly over the years. The NAOMS study would have benefited considerably if it had anticipated the problem in the planning stage. Consultation with other federal agencies (for example, the U.S. Census Bureau) would have avoided many of the problems, both with data release and in losing information content in the released data.

Finding: The issues associated with preserving respondents' anonymity and confidentiality and with the public release of data should have been anticipated and addressed at the design stage of the NAOMS project. There is considerable expertise in this area in both the research literature and among practitioners in the federal agencies. Such advance planning would have avoided the need for after-the-fact, ad hoc redaction methods and the resulting loss of information.

[10] See American Statistical Association, *Privacy and Confidentiality,* 2003.

7

An Assessment of the Utility of NAOMS Data

This chapter continues with task 4 of the charge to the committee requiring that it "conduct an analysis of the NAOMS project survey data provided by NASA to determine its potential utility. . . . This analysis may include an assessment of the data's validity using other known sources of information." Issues on data quality, data validation, and estimation of rates and trends are discussed in this chapter.

7.1 DATA QUALITY

7.1.1 Anomalies

As noted in Chapter 6, NAOMS survey data in the Phase 2 release were not cleaned or modified (unlike those released in Phase 1). One can thus examine the quality of the raw data from the survey. The committee's analysis found that a significant proportion of the non-zero numerical values were implausible, both for event counts (numerators) and for numbers of legs/hours flown (denominators). Selected examples are discussed below.

Table 7.1 shows the distributions of data values for the number of flight legs flown for all pilots who reported that they flew more than 60 flight legs in the 60-day recall period. Data for 3 of the 4 years of the survey (2002 through 2004) are shown separately.[1] Note the high numbers for flight legs flown, with responses as high as 300-650 during the recall period in some cases. Even values in the 150-200 range may be unlikely for a 60-day recall period in an FAR Part 121 operation because of the limitations on pilot flying in regulations and operator scheduling policies. Further, the number of pilots who reported these numbers are not small (15 percent of the pilots reported having flown more than 150 hours). Table 7.2 shows the corresponding distributions for number of hours flown for all pilots who reported that they flew more than 150 hours. Again, note the implausibly high numbers for hours flown and their frequencies, including responses numbering as many as 400-600 hours flown during the recall period.

An equally serious problem exists with event counts. Since many of these events, such as in-flight engine failure, are rare (that is, most of the responses are zero, and only a small fraction of the pilots reported non-zero counts), it is clear that even a few anomalous values can completely distort the estimates of event rates.

---

[1] Because the recall period was not constant during the first year of the NAOMS survey, data for 2001 are excluded from Tables 7.1 and 7.2.

TABLE 7.1 Distributions of the Number of Flight Legs Flown During the 60-Day Recall Period, for the Years 2002-2004

No. of Legs Flown	2002	2003	2004
61-100	944	907	805
101-150	267	209	175
151-200	66	58	42
201-250	13	14	6
251-300	14	9	7
301-350	2	4	2
351-400	2	8	2
401-450	3	1	0
451-500	2	3	2
501-550	1	1	0
551-600	1	0	0
601-650	1	0	0

NOTE: For Category 5: greater than 60 legs.

TABLE 7.2 Distributions of the Number of Hours Flown During the 60-Day Recall Period, for the Years 2002-2004

No. of Legs Flown	2002	2003	2004
151-200	713	716	802
201-250	234	32	18
251-300	19	6	2
301-350	2	2	1
351-400	2	3	1
401-450	2	0	1
451-500	2	0	0
501-550	1	0	0
551-600	1	0	0

NOTE: For Category 6: greater than 150 hours.

The committee's analysis showed that the problem with anomalous values is common across many event types. Table 7.3 provides selected examples. For the event AH4 ("inadvertently landed without clearance at an airport with an active control tower"), a total of 541 events were reported by 161 pilots. Of these, 4 pilots reported 10, 20, 30, and 303 events (the last response corresponding to a pilot who flew between 46 and 70 hours and fewer than 14 legs during the recall period). These 4 pilots accounted for 363 (67%) of the 541 events reported. Table 7.3 shows several other examples with unusually high numbers of events reported. If the instances of such anomalies were limited to only a few event types, one might be able to investigate them in greater detail. Unfortunately, however, the problem was extensive, with one, and often several, implausible values for many event types.

There are at least two possible reasons for these anomalous values: (1) the pilots gave erroneous answers, or (2) errors were made during data entry. A verification question for high values for hours flown was included in the questionnaire, but the committee does not know if other data-audit procedures were in place to flag implausible values reported by the respondents or entered into the database.

7.1.2 Rounding

Another characteristic common in the survey data was the rounding of the responses (raw data) by the respondents (pilots). A disproportionate number of observations were rounded, either to have zero or 5 as the last

TABLE 7.3 Examples of Implausibly High Non-Zero Counts for Events

Survey Question	Total Number of Events	Number of Pilots Who Reported at Least One Event	Unusually High Numbers of Events as Reported by Individual Pilots[a]
AH4: Number of times respondent inadvertently landed without clearance at an airport with an active control tower.	541	161	303, 30, 20, 10. Of the 161 pilots who reported a non-zero count, 4 pilots accounted for 363 of the 541 events (or 67% of the total).
AH8: Number of times respondent experienced a tail strike on landing.	80	24	30, 10(2), 9. Of the 24 pilots who reported a non-zero count, 4 pilots accounted for 59 of the 80 events (or 74% of the total).
AH13: Number of times respondent experienced an unusual attitude for any reason.	508	450	100, 60, 30, 12, 11. Of the 450 pilots who reported a non-zero count, 5 pilots accounted for 213 of the 508 events (or 42% of the total).
ER6: Number of times an in-flight aircraft experienced a precautionary engine shutdown.	365	215	30(3), 20, 10, 9. Of the 215 pilots who reported a non-zero count, 6 pilots accounted for 129 of the 365 events (or 35% of the total).
ER7: Number of times an in-flight aircraft experienced a total engine failure.	132	82	30, 10, 3(2). Of the 82 pilots who reported a non-zero count, 4 pilots accounted for 46 of the 82 events (or 56% of the total).
GE5: Number of times respondent went off the edge of a runway while taking off or landing.	350	33	90, 70, 30(4), 20, 10(2), 9(2). Of the 33 pilots who reported a non-zero count, 11 pilots accounted for 338 of the 350 events (or 97% of the total reported).
GE9: Number of times respondent landed while another aircraft occupied or was crossing the same runway.	928	240	100, 80, 50, 40(2), 20, 15, 12, 10(14). Of the 240 pilots who reported a non-zero count, 20 pilots accounted for 497 of the 928 events (or 54% of the total).

[a] Number in parentheses refers to number of pilots who reported that number of events.

digit. Figure 7.1 shows an example for the number of hours flown in Category 2 (46-70 hours) during the 4-year period of the survey. Similar problems arose with the number of events reported. This type of rounding is common when respondents cannot recall the exact numbers. The committee did not conduct an extensive analysis to assess the magnitude of the rounding bias on the computed event rates. Nevertheless, the distribution of the numbers in Figure 7.1 suggests that it may be significant. This problem could have been alleviated in part by asking the respondent to retrieve his or her logbook to verify the answers. A request along these lines could have been included in the pre-notification letter that was sent to the respondents.

Finding: There are several problems with the quality of NAOMS data:

- **Substantial fractions of the reported non-zero counts of events had implausibly large values, as did the reported flight legs and hours flown. Simple audits to alert for such values should have been used during the computer-assisted telephone interviews and data-cleaning steps to reduce the occurrence of these problems.**
- **It appears that respondents often rounded their answers to convenient numbers (for example, there were unusually high occurrences of numbers with final digits of "0" and "5").**

The extent and magnitude of these problems raise serious concerns about the accuracy and reliability of the data. The development of appropriate strategies for handling some of these problems will require access to the unredacted data.

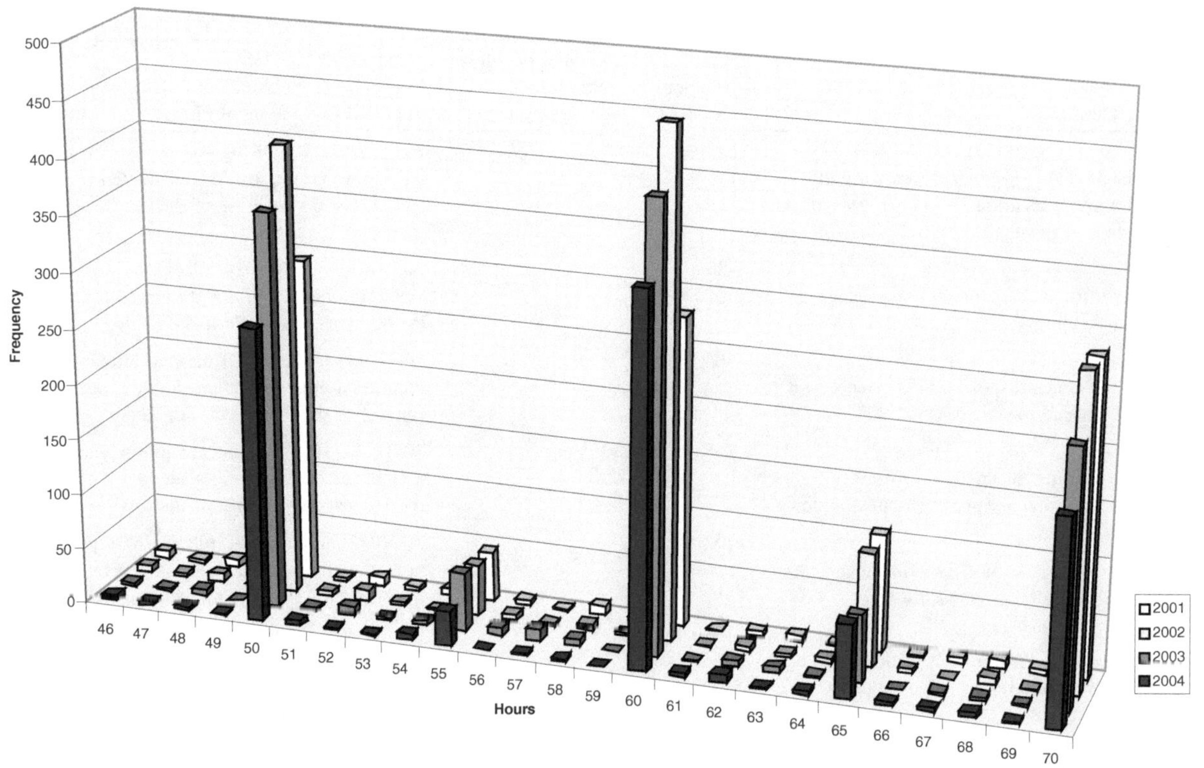

FIGURE 7.1 Rounding of responses for numbers of hours flown in Category 2 (46-70 hours).

7.2 EXTERNAL DATA VALIDATION

7.2.1 Comparisons with Other Data Sources

One type of external validation involves comparing the attributes of the respondents in the sample to corresponding population data from other sources. For example, if the proportion of certain characteristics (distribution of aircraft or distribution of pilots by experience levels) is quite different from the proportion of the same characteristics in another reliable source, the survey results might not be representative. Table 4.1 in Chapter 4 shows that the distribution of the aircraft types in the NAOMS survey differed markedly from that in the BTS data, with the proportion of wide-body aircraft being over-represented in the NAOMS survey.

Similarly, if other data sources were available for event counts, these sources could be used for an external validation of the counts. NAOMS representatives indicated to the committee that they saw no point in asking questions to which answers could be obtained elsewhere. While this is valid point, a limited amount of redundancy is often included in surveys for the purposes of validation. The committee recognizes the potential for problems in comparing data across different sources (differences in contexts, in the way that data were collected, etc.), but such comparisons are often conducted in other surveys and have been extremely valuable.

7.2.2 Use of Logbooks

Another potential source of external validation is the use of respondents' logbooks during the survey. The invitation letter requesting survey participation suggested that respondents have their logbooks readily available during the survey. However, the committee did not find information in the survey or other documents indicating whether the respondents actually referred to their logbooks while answering their questions. The survey could have included a question on this matter, such as, "Are you using your logbook in providing responses during this survey?" This information would have been helpful in assessing the validity of the responses. The response to question D1 in the final section ("How confident are you that you accurately counted all of the safety-related events that I asked you about?") provides a rough measure of a respondent's confidence in the accuracy of the responses, but it is unclear how this information could be incorporated into the estimation process.

Finding: Limited comparison of NAOMS data with those of other sources indicates an over-representation of some groups and an under-representation of others. Such sampling biases must be addressed in the estimation of event rates and trends. More effort should have been spent on ensuring data accuracy at the interview stage, such as by asking respondents to refer to their logbooks. Preliminary analysis of the data would likely have raised these problems in time to modify the survey implementation accordingly.

7.3 ESTIMATION AND WEIGHTING

7.3.1 Overall Rates

Consider the estimation of a particular event type, k, during a given recall period, t, in the AC survey (the issues are similar for the GA survey). Let D_{kt} be the number of events of type k that were observed by all AC pilots during the recall period t. Similarly, let M_t be the total number of flight units (legs or hours as appropriate) flown by all AC pilots during the recall period t. Then, the *true* population rate for event k during period t is

$$R_{kt} = D_{kt}/M_t. \tag{7.1}$$

For example, event k may refer to landings in an occupied runway in this example, and t may denote the time period January 1 through March 31, 2003. In this case, the appropriate denominator (flight units) is the number of flight legs; for other events, it may be the number of flight hours.

Let d_{kt} be the total number of events of type k that were observed in the sample of AC pilots during the recall period t. Similarly, let m_t be the total number of flight units (legs or hours, as appropriate) flown by all AC pilots in the sample during the recall period t. If the survey results are based on a simple random sample (or more generally, an equal-probability design), then the population ratio R_{kt} can be estimated by the corresponding sample ratio

$$r_{kt} = d_{kt}/m_t. \tag{7.2}$$

The properties of this estimate and expressions for its variance under simple random sampling can be found in most textbooks on sample surveys.[2]

However, there are several types of biases present in the NAOMS study that preclude the use of the simple estimate in Equation 7.2. Chapter 4 discussed various types of coverage biases. The over-representation of wide-body aircraft and under-representation of smaller aircraft in the study were noted there. In addition, the sampling probabilities of flight legs varied with the number of pilots in the aircraft, and these unequal probabilities have to be accounted for when estimating event rates.

If there is sufficient information about the precise nature and the magnitude of these biases, it is possible that at least some of them can be accounted for by weighting the responses appropriately. For example, if one knew the unequal sampling probabilities for the flight legs due to the presence of multiple pilots in the aircraft, the responses could be weighted inversely by the sampling probabilities. There is extensive discussion of these methods in the sampling literature.[3] However, this type of information must be documented during the planning and implementation stages of the study and does not appear to be available for the NAOMS survey.

---

[2] See, for example, Cochran, *Sampling Techniques,* 1977.

[3] See, for example, Lohr, *Sampling: Design and Analysis,* 1999.

For the unredacted data, event rates can be computed for a period as small as 2 months (the recall period). For the redacted data, the information is grouped into years, so periods of length of 1 year are the smallest periods for which rates can be calculated. As noted in Chapter 5, this level of categorization severely limits the usefulness of the data. It is difficult to detect any effects because of seasonal variations, short-term effects of changes in aviation procedures on safety data, and other effects likely to be of interest for safety-monitoring purposes.

7.3.2 Rates by Subpopulations

In addition to the overall event rates in Equation 7.1, users of aviation safety data will also be interested in event rates for various subpopulations, such as rates by aircraft size or by pilot experience. Consider, for example, the event "landing on occupied runways," and suppose that one wants to compare how the rate for this event varies by three subpopulations of pilot experience: low, medium, and high levels. Let D_{jkt} be the number of flights that landed in an occupied runway (event type *k*) during the recall period *t* by pilots with experience level *j*. Similarly, let M_{jt} be the total number of flights during the recall period *t* by pilots with experience level *j*. Then, the rate of interest is

$$R_{jkt} = D_{jkt} / M_{jt} . \tag{7.3}$$

Let d_{jkt} be the number of flights that landed in an occupied runway (event type *k*) that were observed in the sample of AC pilots during the recall period *t*. Further, let m_{jt} be the number of flights during the recall period *t* by pilots with experience level *j* in the survey. Then, if the survey results in a simple random sample of pilots and the full data are available, one can estimate the population ratio R_{jkt} by the sample ratio

$$r_{jkt} = d_{jkt} / m_{jt} . \tag{7.4}$$

However, it is not possible to estimate these rates from the redacted data, as the counts d_{jkt} and m_{jt} are not available for subpopulations. As noted for Equation 7.2, the estimates in Equation 7.4 are not valid when there are substantial biases, as appears to be the case with the NAOMS project. Since the nature and extent of the biases were not documented at the planning stage, it was not possible for the committee to examine the use of weighting or other adjustment methods to account for the biases.

Finding: The intended simple random sampling for the NAOMS study would have facilitated the easy computation of event rates. However, the final sample does not appear to be representative of the target population as indicated by the limited data analysis conducted by the committee. The survey sampling literature contains many approaches that attempt to address such coverage problems, but they require detailed information on how the survey was implemented, including the type and nature of problems that were experienced, and access to the original data.

7.3.3 Estimation of Trends

The most consistently articulated goal of the NAOMS project was to use survey data to learn about *trends*. Information on trends allows one to assess the effects of safety innovations on event rates. Preliminary analyses by the NAOMS team appear to indicate that the trends for a number of safety events were consistent over time. However, the committee did not conduct any analysis to verify the results, as it had access only to redacted data, in which the time variable was aggregated to full years.

It is important to recognize that event rate biases discussed thus far in this report would not affect trends *to the extent that the biases are constant over time*. For example, if any biases because of nonresponse were constant across years, those biases would cancel out in estimates of trends. However, some type of biases may not have been constant or may have drifted over the survey period. For example, as discussed in Chapter 5, the AC questionnaire included operations and events from a broad array of aviation industry segments. If the mix of these operations changed over time, this would have caused biases in the trend estimates. In addition, biases associated with subjective assessments by pilots may have changed abruptly in response to external events such as those of September 11, 2001.

Finding: Many of the biases that affect the estimates of event rates may be mitigated for trend analysis to the extent that the biases remain relatively constant over time. However, the degree of mitigation might vary substantially across event types.

7.4 CONFIDENCE INTERVALS

The charge to the committee asked for specific recommendations on how to compute *error bars* for the estimates, or in statistical terminology, *confidence intervals*. The key information needed for computing a confidence interval is the variance of the estimated event rate. Under the equal-probability sampling scheme, the variance of the simple ratio estimate in Equation 7.2 can be computed easily.[4] Given the variance estimate, a normal approximation is generally used to compute the confidence interval. Since these issues have been discussed extensively elsewhere, the committee will not repeat the details here.

The development of confidence intervals (error bars) for the NAOMS study faces the same difficulties that were discussed for the estimates in Section 7.3 and would require knowledge of the nature and extent of the biases that was not available to the committee. Without such information, the committee cannot provide recommendations that will be useful in this particular context.

7.5 SUMMARY

Careful planning of any statistical investigation, including surveys, involves the following steps: (1) the development of initial methods for data analysis and estimation, (2) the analysis of pilot data or early survey data to discover potential problems, and (3) the use of this information to refine the survey methodology. The committee was surprised by the apparent lack of these activities (or at least lack of documentation of these activities). The NAOMS team also did not conduct a formal analysis of the survey data as they became available. This was an especially serious oversight for a project with a research component in which one goal was to learn and to refine the ideas to improve the methodology. In the committee's view, many of the problems that have been identified with the NAOMS survey might well have been detected and corrected if these aspects of the survey planning had been better executed.

Finding: The committee did not find any evidence that the NAOMS team had developed or documented data analysis plans or conducted preliminary analyses as initial data became available in order to identify early problems and refine the survey methodology. This is consistent with any well-conducted research study.

The final charge to the committee asks for recommendations regarding the most effective ways to use the NAOMS data. Because the committee did not have access to the unredacted data, a recommendation on this front, by necessity, only relates to the redacted, publicly available data.

As in any research study, a full description of the NAOMS project and the results of any analysis should be submitted for possible publication, which will involve a peer review. Because of the problems associated with analyzing the redacted data set, discussed in Chapter 6, the analysis would have to be based on the unredacted data and would need to address challenges such as treatment of data errors and potential effects of biases on trends.

However, because of the methodological and implementation problems cited in Chapters 4 and 5 as well as the difficulties associated with data analysis discussed in this chapter, the committee does not recommend using the publicly available data set of NAOMS to identify system-wide trends in the rates of safety-related events.

---

[4] Cochran, *Sampling Techniques,* 1977.

Recommendation: The publicly available NAOMS data should not be used for generating rates or trends in rates of safety-related events in the National Airspace System. The data could, however, be useful in developing a set of lessons learned from the project.

Appendixes

A

Biographical Sketches of the Committee Members

Vijayan N. Nair, *Co-chair,* is the Donald A. Darling Professor of Statistics and a professor of industrial and operations engineering at the University of Michigan. He has been the chair of the statistics department since 1998. Prior to that, he served as a research scientist at Bell Laboratories for 15 years. Dr. Nair's area of expertise is engineering statistics, including quality and productivity improvement, experimental design, reliability, and process control. He is a fellow of the American Association for the Advancement of Science, the American Society for Quality, the American Statistical Association, and the Institute of Mathematical Statistics; an elected member of the International Statistical Institute; and a senior fellow of the Michigan Society of Fellows. Dr. Nair is a former editor of *Technometrics* and the *International Statistical Review* and has served on many other editorial boards. He is currently vice president of the International Statistical Institute, chair of the Statistics Division of the American Society for Quality, and president-elect of the International Society for Business and Industrial Statistics. He is a former chair of the board of trustees of the National Institute of Statistical Sciences. He is a member of the NRC's Board on Mathematical Sciences and Their Applications (BMSA), is serving as the current chair of another NRC Committee, and has served on several other committees. He has a bachelor's degree in economics from the University of Malaya, Malaysia, and a Ph.D. in statistics from the University of California, Berkeley.

Clinton V. Oster, Jr., *Co-chair,* is a professor at the School of Public and Environmental Affairs, Indiana University. His research focuses on air traffic management and aviation infrastructure, with an emphasis on aviation safety. His research also includes airline economics and competition policy. Professor Oster has co-authored four books on aviation safety and policy. He has been a consultant on aviation and other transportation issues to the U.S. Department of Transportation (DOT), the FAA, NASA, the European Bank for Reconstruction and Development, state and local governments, and private-sector companies in the United States, Canada, the United Kingdom, Russia, and Australia. He has participated in several NRC committees, serving as chair for four committees and one expert panel. Professor Oster is a member of the National Aviation Advisory Group of the U.S. Government Accountability Office and of the Aviation Network on Aviation Research and Policy for the Dutch Ministry of Transport. He served as the research director for the Aviation Safety Commission as well as the director of Indiana University's Transportation Research Center and as associate dean at the School of Public and Environmental Affairs. He received a B.S.E. degree in chemical engineering from Princeton University, an M.S. in urban and public affairs from Carnegie Mellon University, and a Ph.D. in economics from Harvard University.

David L. Banks is a professor of the practice of statistics at Duke University. Having worked at NIST, the BTS, and the Food and Drug Administration, Dr. Banks' statistical expertise is amplified by his understanding of some important federal government needs. He recently won the Roger Herriot Award for his unique approaches to the solution of statistical problems in federal data-collection programs. Dr. Banks is credited with pioneering the use of Bayesian statistics for metrology, with helping to build the BTS, and with leading efforts to apply statistical methods for risk analysis and game theory to counter bioterrorism. He has previously served as a member of the NRC's Committee on Applied and Theoretical Statistics (CATS) and is currently a member of BMSA's standing Committee on Biodefense Analysis and Countermeasures. He received his Ph.D. in statistics from Virginia Polytechnic Institute and State University.

Robert M. Bell is a principal member of the technical staff of the Statistics Research Department at AT&T Laboratories. Previously, from 1980 to 1998, Dr. Bell was associate statistician, statistician, and senior statistician at the RAND Corporation. He also worked as the head of the RAND statistics group from 1993 to 1995. He is an expert in experimental design and survey development, data analysis, and statistical methodology. Dr. Bell recently served as chair of the NRC's Panel on Correlation Bias and Coverage Measurement in the 2010 Census. He has previously served as a member on five other NRC committees, including the Committee on National Statistics, and served as chair of the Committee to Review the 2000 Decade Design of the Scientists and Engineers Statistical Data System. He is a fellow of the American Statistical Association (ASA) and has also served as member and chair of the ASA Census Advisory Committee of Professional Associations. He received his B.S. in mathematics from Harvey Mudd College in 1972, his M.S. in statistics from the University of Chicago in 1973, and his Ph.D. in statistics in 1980 from Stanford University.

Johnny Blair is a senior survey methodologist and principal scientist at Abt Associates. He has conducted methodological research in a number of areas, including sampling from rare populations, measurement error in proxy reporting, data quality in converted refusal interviews, the use of incentives to increase response rates, and, most recently, the design and analysis of cognitive interviews for pretesting. Mr. Blair is co-author of the book *Designing Surveys: A Guide to Decisions and Procedures.* From 2001 to 2006, he served on the ASA Committee on Energy Statistics, which advises the Energy Information Administration in the establishment of survey design and instrument development. He is also a member of the Design and Analysis Advisory Committee to the Educational Testing Service for the National Assessment of Educational Progress.

Anthony J. Broderick is an independent aviation safety consultant who works with domestic and international airlines, aerospace firms, a major aircraft manufacturer, and governments. Before retiring from his post as associate administrator for regulation and certification in the FAA, Mr. Broderick served for 11 years as a senior aviation safety official in the U.S. government. He led the FAA's development of the International Aviation Safety Assessment program and was also instrumental in leading international efforts to establish certification and operational standards for safety. Prior to this appointment, Mr. Broderick spent 14 years at the FAA and the DOT and 7 years in private industry. His portfolio also includes a background in civil aviation security; aviation environmental issues; management of the FAA evaluation, currency, and transportation flying programs; and oversight of the FAA flight inspection program. Mr. Broderick is a private pilot. He has received many awards and recognition for his work in the aeronautics industry and is a fellow of the Royal Aeronautical Society. He was previously a member of the Aeronautics and Space Engineering Board and has served on three other NRC studies: the Panel on Transportation for Science and Technology for Countering Terrorism, the Committee on Aeronautics Research and Technology for Environmental Compatibility, and the Committee to Conduct an Independent Assessment of the Nation's Wake Turbulence Research and Development Program.

James Danaher retired in 1998 as chief of the Operational Factors Division of the Office of Aviation Safety, National Transportation Safety Board (NTSB). He has more than 35 years of government and industry experience in the human factors and safety fields. After joining NTSB in 1970, he served in various management positions, with a special emphasis on human performance in flight operations and air traffic control. Mr. Danaher has participated in the on-scene investigation of numerous accidents, in public hearings, and in the development of NTSB recommendations. He is a former naval aviator and holds a commercial pilot's license with single-engine, multi-engine, and instrument ratings. Among other NRC assignments, he has served on the Committee for the Review of NASA's Revolutionize Aviation Program (a committee that had interactions with the NAOMS project). Mr. Danaher holds a master's degree in experimental psychology from Ohio State University.

Peter Griffiths is the regional vice president for Europe for the International Air Transport Association. He was previously director general of the United Kingdom Civil Aviation Authority, a public corporation with the responsibility of overseeing and regulating all aspects of aviation in the United Kingdom. Mr. Griffiths joined the aviation industry in 1998 as a first officer flying a Boeing 737 aircraft for Go, a British Airways subsidiary, and went on to be a captain on Airbus aircraft. He was the security manager and then director of safety and security for easyJet, a low-cost airline based at London's Luton Airport. Mr. Griffiths is a trained engineer; has a master's degree in risk, crisis, and disaster management; and has qualifications in air accident investigation.

Iain M. Johnstone is a professor in the Departments of Statistics and of Health Research and Policy (Biostatistics) at Stanford University. His research interests include statistical decision theory, wavelet-like methods in estimation theory, and multivariate analysis. Dr. Johnstone was elected to the National Academy of Sciences in 2005 for his fundamental contributions to the understanding of statistical procedure for the analysis of the enormously complex and multidimensional data that are arising in many fields. He previously served on BMSA. He holds an M.Sc. in probability and statistics and a B.Sc. in pure mathematics and statistics from Australian National University, and an M.S. and a Ph.D. in statistics from Cornell University.

Karen Kafadar is the James H. Rudy Professor of Statistics and Physics at Indiana University, Bloomington. She received her B.S. and M.S. degrees from Stanford University and her Ph.D. in statistics from Princeton University. Her research focuses on exploratory data analysis, robust methods, characterization of uncertainty in quantitative studies, and analysis of experimental data in the physical, chemical, biological, and engineering sciences. Prior to joining Indiana University, she was a professor and Chancellor's Scholar in the Departments of Mathematical Sciences and of Preventive Medicine and Biometrics at the University of Colorado-Denver, a fellow at the National Cancer Institute (cancer screening section), and a mathematical statistician at Hewlett Packard Company (research and development laboratory for radio-frequency/microwave test equipment) and at the National Institute of Standards and Technology (where she continues as a guest faculty visitor on problems of measurement accuracy, experimental design, and data analysis). Dr. Kafadar's previous engagements include consultancies in industry and government as well as visiting appointments at the University of Bath, Virginia Polytechnic Institute, and Iowa State University. She has served on previous NRC committees and chaired CATS. She also serves on the editorial boards of several professional journals as editor or associate editor and on the governing boards of the ASA and the Institute of Mathematical Statistics. She is a fellow of the ASA and an elected member of the International Statistical Institute, has authored more than 90 journal articles and book chapters, and has advised numerous M.S. and Ph.D. students.

Elizabeth A. Lyall has served as a consultant and contractor to the FAA on issues related to human factors issues of flight deck automation design, training, operations, and certification for more than 15 years. She founded Research Integrations, Inc., to be an independent voice for influencing flight safety by conducting and applying relevant research. She has served as a member of the International Harmonization Working Group to develop a new regulation addressing human factors in flight deck design for transport-category airplanes and is currently serving on the Flight Deck Automation Working Group that is identifying current and future safety and other operational issues with the design, training, operation, and certification of flight deck automated systems and their interaction in the current and future airspace. Research Integrations also has developed and maintains the website www.flightdeckautomation.com, which includes a searchable database of flight deck automation issues and related research findings for each of those issues. Dr. Lyall received her Ph.D. from Arizona State University.

Donald W. Richardson retired in 2005 as the vice president of Science Applications International Corporation (SAIC) responsible for all FAA and civil aviation corporate activities. He has been an active pilot for 59 years and possesses a breadth of experience with multi-engine land and seaplanes. His engineering career has included assignments as an aerodynamics and flight test engineer, research pilot, and engineering manager. He is a fellow and past president of the American Institute of Aeronautics and Astronautics (AIAA), of which he has been a member for 58 years. He is also a fellow of the Royal Aeronautical Society and has served on its Engineering Council. He was awarded the NASA Public Service Medal in 2002 for his work in reinvigorating U.S. federal funding in research and development in aeronautics. He holds B.S., M.S., and Ph.D. degrees in aeronautical and mechanical engineering.

Thomas B. Sheridan is Ford Professor of Engineering and Applied Psychology Emeritus in the Department of Mechanical Engineering and Department of Aeronautics and Astronautics at the Massachusetts Institute of Technology (MIT) and is senior transportation fellow at the DOT Volpe National Transportation Systems Center. Dr. Sheridan's research interests include modeling and experiment on human-automation interaction in aviation, highway, medical, and other systems. He is the author or co-author of five books and more than 200 scientific papers. He served as president of the IEEE Systems, Man and Cybernetics Society and as editor of *IEEE Transactions on Man-Machine Systems;* he received the IEEE's Norbert Wiener and Joseph Wohl awards and the IEEE Centennial and Millennium Medals; and he is an IEEE fellow. Dr. Sheridan is also a fellow of the International Ergonomics Association and the Human Factors and Ergonomics Society (HFES), a recipient of several HFES awards, and a past president of HFES. He received the National Engineering Award of the American Association of Engineering Societies and the Oldenburger Medal of the American Society of Mechanical Engineers. Dr. Sheridan has served as a member of several NRC committees, including the Committee for the Review of NASA's Revolutionize Aviation Program (a committee that had interactions with the NAOMS project). He is a member of the National Academy of Engineering. Dr. Sheridan received his B.S. degree in mechanical engineering from Purdue University, an M.S. degree in engineering from the University of California at Los Angeles, an Sc.D. degree from MIT, and an honorary doctorate from Delft University of Technology, Netherlands.

Alfred T. Spain retired as senior vice president of operations at JetBlue Airways Corporation in May 2006. Previously, he had served in various capacities at Continental Airlines, including as vice president of flight operations for Continental Micronesia, Inc. Mr. Spain is pilot-rated in numerous multi-engine and turbine-powered aircraft. Previously he had been a commercial pilot and instructor for more than 20 years. He is a senior member of the AIAA, a life member of the Navy League of the United States, and a member of the Aircraft Owners and Pilots Association and the Seaplane Pilots Association. He has experience in the senior management of flight operations for domestic and international airlines as well as civilian and military experience in safety applications for both flight and ground operations. Mr. Spain has a B.S. in professional aviation from Louisiana Technical University and an M.B.A. from Concordia University.

S. Lynne Stokes is a professor of statistical science at Southern Methodist University in Dallas, Texas. Her current research interests are sampling methods, modeling of nonsampling errors in surveys, and disclosure limitation methods. Before working at Southern Methodist University, Dr. Stokes was a mathematical statistician at the Patuxent Wildlife Research Center for the U.S. Fish and Wildlife Service, at the Center for Social Science Research at the U.S. Bureau of the Census, and as faculty at the University of Texas at Austin. She is a fellow of the ASA and the past chair of the ASA's Survey Research Methods Section. Dr. Stokes currently serves as the associate editor for *Survey Methodology* and has served as editor of *The American Statistician.* She has served on the NRC Committee on Review of Recreational Fisheries Survey Methods and on the Panel on Alternative Census Methodologies. She received her Ph.D. in statistics from the University of North Carolina at Chapel Hill.

B

Individual Presenters to the Committee

Michael Basehore, ASIAS program manager, Federal Aviation Administration
Thomas Chidester, formerly Manager, Human Factors and Safety Training, American Airlines
Renato Colantonio, NAOMS 2008 Project Manager, NASA Glenn Research Center
Linda Connell, Director, NASA Aviation Safety Reporting System, NASA Ames Research Center
Mary Connors, Chief, Aviation System Safety Research Branch, NASA Ames Research Center
Robert Dodd, Principal Investigator of NAOMS, Battelle Memorial Institute
Vernon Ellingstad, Director, Office of Research and Engineering, National Transportation Safety Board
Keith Hagy, Director of Engineering and Air Safety, Air Line Pilots Association
Jon Krosnick, Frederic O. Glover Professor in Humanities and Social Sciences, professor of communication, and professor of political science, Stanford University
Bruce Landsberg, Executive Director, Aircraft Owners and Pilots Association
Bryan O'Connor, Associate Administrator, Office of Safety and Mission Assurance, NASA
Steve Predmore, Vice President and Chief Safety Officer, JetBlue
Loren Rosenthal, Manager, Battelle Mountain View Operations, Battelle Memorial Institute
Jaiwon Shin, Associate Administrator, Aeronautics Research Mission Directorate, NASA
Irving Statler, Ames Associate, NASA Ames Research Center
Simon Stewart, Safety Management System Development and Training Manager, EasyJet

C

Acronyms

AC	air carrier
ACD	Airmen Certification Database
AIDS	Accident/Incident Data System
ALPA	Air Line Pilots Association
AQP	Advanced Qualification Program
AR	Air Registry
ASA	American Statistical Association
ASAP	Aviation Safety Action Program
ASEB	Aeronautics and Space Engineering Board
ASIAS	Aviation Safety Information Analysis and Sharing System
ASMM	Aviation System Monitoring and Modeling Project
ASRS	Aviation Safety Reporting System
ATC	Air Traffic Control
ATOS	Air Transportation Oversight System
ATP	air transport pilot
AvSP	NASA's Aviation Safety Program
BTS	Bureau of Transportation Statistics
CAST	Commercial Aviation Safety Team
CATI	computer-assisted telephone interview
CATS	Committee on Applied and Theoretical Statistics
FAA	Federal Aviation Administration
FAR	Federal Aviation Regulations
FE	flight engineer
FOIA	Freedom of Information Act
FOQA	Flight Operational Quality Assurance

GA	general aviation
GAIN	Global Analysis and Information Network
JIMDAT	Joint Implementation Measurement Data Analysis Team
NAEP	National Assessment of Educational Progress
NAOMS	National Aviation Operations Monitoring Service
NAS	National Airspace System
NASA	National Aeronautics and Space Administration
NCARC	National Civil Aviation Review Commission
NCS	National Crime Survey
NCVS	National Crime Victimization Survey
NIBRS	National Incident Based Reporting System
NMACS	Near Midair Collision System
NRC	National Research Council
NTSB	National Transportation Safety Board
UCR	Uniform Crime Report
WAAS	World Aviation Accident Summary

D

Principal Segments of Aviation in the United States

Title 14 of the *Code of Federal Regulations* (CFR) governs Aeronautics and Space. Within Title 14, four parts are of interest in understanding the segments of aviation addressed in this report. Those parts—91, 119, 121, and 135—are described below.

Part 91 covers "General Operating and Flight Rules" and prescribes rules governing the operation of aircraft within the United States. For persons operating civil aircraft as air carriers or commercial operations, Part 119 prescribes the type of operating certificate that must be obtained, with accompanying additional regulations, specifically whether operations must be under Part 121 or Part 135. Part 121 prescribes the regulations, beyond those prescribed in Part 91, that must be followed for large aircraft operations and for scheduled operations in smaller aircraft capable of carrying 10 or more passengers. Part 135 prescribes the regulations, again beyond those prescribed in Part 91, for scheduled commuter air carriers using aircraft capable of carrying nine or fewer passengers and for on-demand, for-hire air taxi and charter operations carrying passengers and/or freight. Title 14 of the CFR has several other parts that govern specialized air operations such as agricultural aircraft operations and National Parks Air Tours, but those operations did not play a significant role in the NAOMS survey and are not described here.

Flight operations are often referred to by the part of the regulations that governs them. So, large operators such as American Airlines, Delta Air Lines, Federal Express, Southwest Airlines, and United Airlines are typically referred to as Part 121 carriers. Similarly, small scheduled commuter operators, on-demand air taxis, and small charter freight operators are typically referred to as Part 135 carriers. Finally, general aviation operations by private pilots and corporate flight operations are typically referred to as Part 91 operations.

As discussed in the body of the report, the committee was concerned about the aggregation of these different segments of aviation in the NAOMS survey. One of the impacts of this aggregation was the very large number of flights reported by some of the respondents in the AC survey. As can be seen in Table D.1, the number of hours that a pilot is permitted to fly during specific time periods is different in the different segments of aviation. Note that there are no limitations on pilot flight time in Part 91 operations and that the limitations are different in the scheduled and the nonscheduled segments of Part 135. In addition, other characteristics vary across the types of operations, and these variations could impact the interpretation of results if they are aggregated, such as typical frequency-of-flying and pilot-training requirements.

The committee was also concerned about this aggregation because historically, the safety performance of these different segments of aviation has been notably different, as Table D.2 shows. By combining segments of aviation with disparate safety performance in the NAOMS survey, the opportunity to gain insight into the causes of the differences in safety performance was lost.

TABLE D.1 Limitations on Pilot Flight Time (hours) by Category of Operation

Category	During 7 Consecutive Days	In a Calendar Month	In a Calendar Quarter or 90-Day Period	In a Calendar Year
Part 121 one- or two-pilot crews	32	100		1,000
Part 121 two pilots plus additional flight crew member		120	300	1,000
Part 135 scheduled	34	120		1,200
Part 135 nonscheduled			500	1,400
Part 91	No limit	No limit	No limit	No limit

SOURCE: 14 CFR 121.481, 14 CFR 121.483, 14 CFR 135.265, 14 CFR 135.267.

TABLE D.2 Fatal Accident Rates, by Category of Operation, 1983-2008

Category	Fatal Accidents	Flight Hours	Fatal Accidents per 100,000 Flight Hours
Part 121 scheduled	64	357,075,443	0.018
Part 121 nonscheduled	25	16,915,795	0.148
Part 135 scheduled	78	35,201,228	0.222
Part 135 nonscheduled	575	77,989,000	0.737
Part 91	10,250	674,227,000	1.520

SOURCE: Available at http://www.ntsb.gov/aviation/Stats.htm.

E

Additional Examples of Surveys by Federal Agencies

This appendix provides a brief overview of the practices and experiences of a few illustrative federal surveys. It is intended to show that (1) the surveys have tackled and adequately dealt with survey methodological and operations management issues that were faced by NAOMS; and (2) often, repeated efforts over time, with sufficient resources, are needed before the survey can provide good-quality results. In fact, a first-rate federal survey that still uses its original design and data-collection procedures is rare. It is far more common that, even though initial efforts may be quite competent, excellence is attained over time. At different points during their evolution, these surveys gradually improve, but also sometimes—for example, in the case of the National Assessment of Educational Progress (NAEP) discussed below and the NCVS discussed in Chapter 3—undergo major design changes. Such changes have to be preceded by careful development and testing before implementation.

Another feature of successful government surveys is that they typically have a research component to support the investigation of issues or problems of particular import. Each of them also has a core staff dedicated to the survey's ongoing improvement and adaptation to change. Put another way, these major government survey programs develop a professional organizational culture that fosters the approaches noted and also produces professional staff in both technical and administrative areas that are very knowledgeable about the particular survey's issues and history.

NATIONAL ASSESSMENT OF EDUCATIONAL PROGRESS

The NAEP is an educational assessment that is administered to a probability sample of schoolchildren across the United States. It is conducted by contractors under the direction of the National Center for Education Statistics. It was initiated in 1969 to measure the performance of U.S. schoolchildren in mathematics, reading, and other subjects. The NAEP was established even though there were other ongoing student assessments at the district and state levels. The existence of these other data sources caused some to question the value of the NAEP and raised concerns about whether NAEP results may disagree with other data. In the end, it was determined that accurate evaluations of the effect of educational policies and change over time could be made only through an assessment of the entirety of the educational system such as through the NAEP, and not through a self-selected sample that the available testing programs provided.

The students are tested in each subject using sophisticated test conceptualization, test construction, administration, and scoring. The NAEP requires the combined and coordinated expertise of educators, statisticians, psychometricians, survey sampling statisticians, and methodologists. While the design and measurement processes are complex, the results must be reported in a manner that is useful to policy makers and researchers, yet comprehensible at some level to the general population of parents and teachers.

This complex system of multiple ongoing surveys, with its demand for high levels of both accuracy and timeliness, did not emerge full-blown on the first attempt. Over the four decades of its existence, almost every technical aspect of design and implementation has undergone revision while still, with some exceptions, maintaining trend measures of performance changes over time.

The NAEP provides several lessons for NAOMS. The nature of the concepts that the NAEP measures has changed over time. For example, the content of mathematics topics and skills in 1970 are vastly different from those in 2000. The availability of alternative performance measures based on other tests has also continued over this time, with the introduction of more state testing programs. The NAEP has evolved over time to accommodate these changes, while still regularly producing reliable data that allow trend estimation.

BEHAVIOR RISK FACTOR SURVEILLANCE SYSTEM

The Behavior Risk Factor Surveillance System (BRFSS), started in 1984, is managed by the Centers for Disease Control and Prevention, and is another example of a survey designed to identify trends over time. It is designed to collect information about health behaviors (known to be related to particular health risks) of the general population of the United States. Unlike the NAEP, which is entirely a centralized enterprise, the BRFSS is designed and managed by the individual states, with federal guidelines. This approach came about for historical and political reasons, as well as for technical and resource considerations.

The BRFSS is a telephone survey of the general population that is repeated annually (through the aggregation of quarterly data), using a core instrument that is supplemented with state-selected questions. From a sample design perspective, the BRFSS has to be concerned with issues of frame coverage. The sample design omits cellphone-only households and, by definition, nontelephone households. These frame shortcomings are not stationary "targets," but change over time. The cellphone-only households are increasing and are known to have generally younger residents. The number of nontelephone households varies somewhat with the economy, and the group is known to be disproportionately likely to contain young children. The estimation procedures try to extend results, in a defensible manner, to the nontelephone household population.

The survey relies on respondent self-reports of behaviors within a specified reference period. Some of the behaviors (for example, substance abuse) are socially undesirable and thus potentially subject to under-reporting. All of the behaviors are also subject to errors of memory and frequency. These various challenges have necessitated a program of methodological research. This research has included consideration of alternative estimators for extending results to the nontelephone population and of questionnaire design to find out how best to encourage the reporting of socially undesirable behaviors.

2001 NATIONAL HOUSEHOLD TRAVEL SURVEY

The National Household Travel Survey (NHTS), conducted for the Bureau of Transportation Statistics (BTS), measures both the daily and the long-distance travel of the U.S. household population. The NHTS is an example of a survey designed to update and build on information from prior survey series. The earlier surveys focused on either daily travel (Nationwide Personal Transportation Survey) or long-distance trips (American Travel Survey). The NHTS was intended to cover both, as well as walking and biking trips.

The survey was conducted from March 2001 through May 2002. Households were recruited by telephone. The recruited households were each sent a survey form and asked to report all travel by household members on a randomly assigned "travel day." Telephone interviewers conducted a follow-up interview that asked respondents about their travel on the travel day and the preceding 27 days.

This survey has a number of design issues relevant to NAOMS. It required self-reports of travel, some requiring very detailed information, over a previous reference period. The survey also illustrates one way to deal with a major public event that occurs during the data-collection period and is clearly related to the survey measurements. The occurrence of such an event obviously presents challenges to satisfying the survey's original goals. Such an event also adds a goal—to measure the impact of the unexpected event—that the survey was not intended to accommodate. In the case of the NHTS, the pre- and post-September 11 data were reweighted to make each group a nationally representative sample. Rather than using the exact event day (9/11), the groups were defined by a September 25, 2001, cutoff (though, clearly, some long-distance trips could not logically be divided by that exact date).

In addition to providing a case study of one way to deal with a major, unanticipated event, the mere possibility of such an event may have implications for the design of a survey such as NAOMS, perhaps supporting the use of relatively small sample replicates that can be more easily combined for estimates preceding and following a particular date than can a single annual sample. The altering of the sample design to accommodate an event that in most instances will not occur has to be balanced against the effect of the design change on other survey goals as well as on the cost of operations.

The NHTS illustrates a sort of hybrid design, in which the survey continues a previous survey series, in the sense of estimation objectives, but is not a replication of the previous surveys. The NHTS provides an example of how a survey series that proceeds intermittently might take advantage of what has been learned about the design challenges, extend the data series, and also add new measures.

SURVEY OF RESPIRATOR USE AND PRACTICES

The Survey of Respirator Use and Practices was an establishment mail survey conducted by a contractor for the Bureau of Labor Statistics (BLS) in 2001. The objective was to obtain data that would permit an assessment of compliance with regulations and safe practices, as well as an estimation of the types of equipment in use under particular conditions and by different employee groups.

The sample of establishments, stratified by type and size of firm, was selected from a BLS database. The identification of establishments eligible for the survey was based on information that these establishments had provided to BLS in an earlier, unrelated survey. It was unknown exactly which companies used respirators, but this prior survey information was considered a predictor of respirator use at a company. One effect of this consideration was to carry over nonresponses from the previous survey into the current one. Determining how fully the actual target population was covered was also problematic.

The questionnaire was designed by survey sponsors and pretested with cognitive interviews and a field test. Data collection was by mail addressed to a "knowledgeable person" at each establishment. This essentially first-time federal survey followed standard practices but still encountered a number of problems, some of them very similar to those identified in the NAOMS survey.

The following excerpts of findings from an NRC report summarizing a review of the survey[1] illustrate that first-time surveys often encounter similar problems, which while not "fatal," need to be corrected and can be corrected if the opportunity for subsequent waves of data collection are available:

- The survey was an important first step in collecting respiratory protection data from a probability sample.
- There was insufficient documentation and detail . . . to reconstruct key aspects of the methodology.
- The field test paid little attention to exploring validation procedures that might have provided information on the quality of data.
- There were several material weaknesses in the procedures for instrument testing.
- NIOSH [The National Institute for Occupational Safety and Health] did not set specific precision guidelines for key estimates.

These findings were followed by a series of recommendations to address the listed problems and other flaws should a follow-up survey be conducted at some future date.

[1] National Research Council, *Measuring Respirator Use in the Workplace*, William D. Kalsbeek, Thomas J. Plewes, and Ericka McGowan, eds., The National Academies Press, Washington, D.C., 2007.

REFERENCES

FAA (Federal Aviation Administration). 1995. Aviation Safety Action Plan: Zero Accidents: A Shared Responsibility. Washington, D.C., February 9. Available at http://www.dtic.mil/cgi-bin/GetTRDoc?AD=ADA307192&Location=U2&doc=GetTRDoc.pdf. Accessed July 15, 2009.

FSF (Flight Safety Foundation). 2009. Global Aviation Information Network (GAIN). Alexandria, Va. Available at http://www.flightsafety.org/gain_home.html. Accessed July 15, 2009.

GAO (Government Accountability Office). 2000. Aviation Safety: Safer Skies Initiative Has Taken Initial Steps to Reduce Accident Rates by 2007. RCED-00-111. Washington, D.C. June 30.

NCARC (National Civil Aviation Review Commission). 1997. Avoiding Aviation Gridlock and Reducing the Accident Rate, Final Report. Washington, D.C. December. Excerpt available at http://www.faa.gov/library/reports/aviation/final/media/NCARC_Rpt_PartIII_SectIII.pdf. Accessed July 15, 2009.

F

List of Additional Survey Questions with Problems Discussed in Chapter 5

This appendix lists additional questions from the AC and GA questionnaires of the NAOMS survey that were identified by the committee as potentially problematic.

> AC2. During the last (TIME PERIOD), how many times did an aircraft on which you were a crewmember perform an evasive action to avoid an imminent in-flight collision with another aircraft that was never closer than 500 feet including evasive action in response to a TCAS advisory? [Question GAC2 on the GA questionnaire asks for the same information.]

In addition to the complex structure pointed out in Section 5.2.3, question AC2 asks the pilot if the aircraft was or was not closer than 500 feet during an evasive maneuver. It would be difficult for pilots to make such an assessment, so the question is likely to result in answers that are not very reliable.

> AC3. During the last (TIME PERIOD), how many times did an aircraft on which you were a crewmember experience less than 500 feet of separation from another aircraft while both aircraft were airborne? [Question GAC3 on the GA questionnaire asks for the same information.]

The concern noted in the first sentence for question AC2 above also applies here, particularly if either aircraft had taken an evasive maneuver.

> ER1. How many times during the last (TIME PERIOD) did an aircraft on which you were a crewmember divert to an alternate airport or return to land because of an aircraft equipment problem? A. What systems caused the diversion or return to land? [Question GER1 on the GA questionnaire asks for the same information.]

Without post-flight analysis data, the pilot may not have known the actual cause of a failure of any specific system. For example, a pilot could easily have thought it was a hydraulic system failure when it actually was an electrical pump failure.

> ER2. How many times during the last (TIME PERIOD) did an aircraft on which you were a crewmember experience a spill, fire, fumes, or aircraft damage due to transporting hazardous materials? C. (How many of these times were the spills, fire, fumes or aircraft damage/Was the spill, fire, fumes or aircraft damage) caused because the hazardous materials in question were out of compliance with regulations?

Although an air carrier pilot is required to know the regulations for "loading" hazardous material on board the aircraft, the pilot is not required to know and would not likely have known the specific regulations for packaging in order to determine that the material was not packaged in accordance with the regulations.

ER4. How many times during the last (TIME PERIOD) did an in-flight aircraft on which you were a crewmember experience uncommanded movements of any of the following devices...? [Question GER2 parts A through F on the GA questionnaire asks for the same information].

Without a post-flight analysis of the data, the pilot is unlikely to know which devices experienced uncommanded movements. Specifically, the pilot would not necessarily know if one surface moved initially and thus caused another or other surfaces to move, the secondary movement being what was observed and reported.

ER7. During the last (TIME PERIOD) how many times did an inflight aircraft on which you were a crewmember experience a total engine failure? [Question GER5 on the GA questionnaire asks the same question.]

A situation that might be interpreted as a total engine failure by the pilot may not, in the final analysis, be considered or reported as an engine failure in FAA or industry databases. For example, an accessory or component failure that reduced thrust or revolutions per minute could cause the perception of an engine failure, but upon analysis, it would be determined that the engine itself did not fail and the event would not be reported as an engine failure in FAA or industry databases because the event did not meet the established definition of an engine failure.

TU2. During the last (TIME PERIOD), how many times did an aircraft on which you were a crewmember encounter wake turbulence that resulted in 10 or more degrees of aircraft roll? [Question GTU2 on the GA questionnaire asks a similar question.]

The actual extent of the roll cannot be reasonably determined by a pilot without post-flight analysis of the flight data recorder. At some airlines, such post-flight analysis has often indicated that the "startle factor" of uncommanded pitch, roll, or yaw change has resulted in pilots overestimating the degree of change encountered. General aviation planes are not typically equipped with flight data recorders, which rules out even the possibility of a post-flight analysis of data in this segment.

WE5. During the last (TIME PERIOD), how many times did an aircraft on which you were a crewmember encounter wind shear or a microburst condition that resulted in an airspeed deviation of 15 knots or greater? [Question GWE4 on the GA questionnaire asks for the same information.]

The extent of the airspeed deviation cannot be verified without post-flight analysis of the flight data recorder. In such atmospheric conditions, it is very difficult for the pilot to assess the airspeed deviation accurately because of the limited time and ability to view the airspeed indicator during a recovery procedure.

There is an additional question—GER7, below—on the GA questionnaire that the committee believes asks pilots for information that they were unlikely to be in a position to have.

GER7. During the last 60 days when you were a pilot or copilot, how many times did you discover that (an airplane/a helicopter) had incorrect or bogus parts installed?

A pilot would not likely know that an incorrect or bogus part had been installed unless the people performing maintenance or repair on the aircraft communicated that information to the pilot.

The following questions from the AC and GA questionnaires were identified by the committee as questions in which there were problems with the structure and/or wording of the question:

AH10. During the last (TIME PERIOD), how many times did an aircraft on which you were a crewmember take off with an out-of-limit center of gravity?

It is unclear if question AH10 asks whether the pilot knew about the center of gravity (CG) condition prior to starting the takeoff roll or whether the pilot did not know about the CG condition until becoming airborne and experiencing flight control surface difficulties. In the first case, the error is with the pilot's judgment for taking off in an aircraft known to have an out-of-limit CG, and in the second case the error is with the ground crew for failure to load the aircraft within its CG limits. From the standpoint of trying to reduce such potentially unsafe events, distinguishing between these two markedly different causes would be critical.

AH11. During the last (TIME PERIOD), how many times did an aircraft on which you were a crewmember take-off overweight?

It is unclear if the question asks whether the pilot knew about the overweight condition prior to starting the takeoff roll or if the pilot did not know about the overweight condition until becoming airborne and experiencing aircraft performance difficulties. In the first case, the error is with the pilot's judgment for taking off in an aircraft known to be overweight, and in the second case the error is with the ground crew for failure to load the aircraft within its weight limits. From the standpoint of trying to reduce such potentially unsafe events, distinguishing between these two markedly different causes would be critical.

AH13. During the last (TIME PERIOD), how many times did an aircraft on which you were a crewmember experience an unusual attitude for any reason?

There is no clear definition of what constitutes "unusual" attitude, and the perception of "unusual" is likely to vary considerably among pilots and aircraft types/models. As noted in Section 5.2.3, cognitive testing of such questions should have been conducted during the questionnaire design stage.

AH15. During the last (TIME PERIOD), how many times did an aircraft on which you were a crewmember nearly collide with terrain or a ground obstruction while airborne?

The term "nearly collide" can be interpreted broadly. Some pilots might not consider an event a near collision with terrain unless the Ground Proximity Warning System (GPWS) is activated. Other pilots might apply different criteria or be in aircraft not equipped with a GPWS.

AT1. During the last (TIME PERIOD), how many times was an aircraft on which you were a crewmember unable to communicate with A.T.C. in a time-critical situation because of frequency congestion?

The interpretation of "time critical" is likely to vary across respondents. Some pilots would report "time critical" when they could not communicate with ground control or clearance delivery even though the aircraft was not underway at the time, whereas others would not. Again, cognitive testing of the question would have been appropriate.

AT2. How many times during the last (TIME PERIOD) did an aircraft on which you were a crewmember fly at an undesirably high altitude or airspeed on approach due to an A.T.C. clearance?

What a pilot might consider an "undesirably high altitude or airspeed on approach" is likely to vary significantly with the type of aircraft being flown. This question also combines two different events, high altitude and high airspeed, into a single response.

GE2. During the last (TIME PERIOD), how many times did an aircraft on which you were a crewmember collide or nearly collide with a ground vehicle?

As with Question AH15, the term *nearly* can be interpreted broadly. Parts A through C of this question ask where the *near collisions* occurred but provide no definition of *near collision*. The interpretation of *nearly* or *near collision* is likely to vary among respondents and circumstances. For example, if one pilot saw the ground vehicle and could determine that its path was not going to result in a collision even though it would pass close to the aircraft, that pilot might not consider that a near collision. However, if another pilot did not see the ground vehicle until the last minute when it passed close to the aircraft, that pilot might be startled and consider the event a near collision even though the vehicle might not have been any closer to the aircraft in the second situation than in the first.

GE3. During the last (TIME PERIOD), how many times did an aircraft on which you were a crewmember skid, slide, or hydroplane resulting in a significant increase in stopping distance during landing?

There is often some degree of hydroplaning resulting in some increase in stopping distance when landing on a wet runway, particularly one with puddles of standing water. A pilot's perception of what was significant could well vary depending on the circumstances. For example, an increase in stopping distance might be considered significant on a short runway, whereas the same increase in stopping distance might not be considered significant on a long runway.

GE10. During the last (TIME PERIOD), how many times did an aircraft on which you were a crewmember nearly

> experience a ground collision with another aircraft while both aircraft were on the ground?

As with question GE2, this question provides no definition of what constitutes *nearly.* Parts A through C of this question ask where the *near collisions* occurred, but provide no definition of *near collision.* Refer to the explanation above, provided for question GE2.

> TU1. During the last (TIME PERIOD), how many times did an aircraft on which you were a crewmember encounter severe turbulence that caused large abrupt changes in altitude, airspeed, or attitude? [Question GTU1 on the GA questionnaire asks for the same information.]

The interpretations of the terms *severe turbulence* or *large abrupt changes in altitude, airspeed, or attitude* can vary considerably. Even if definitions had been provided, it is unlikely that the extent of the turbulence or change in altitude, airspeed, or attitude could be verified with needed accuracy without post-flight analysis of the flight data recorder. In addition, question TU1 asks the respondent to combine multiple events into a single response.

> WE1. During the last (TIME PERIOD), how many times did an aircraft on which you were a crewmember lack accurate weather information when crewmembers needed it while airborne? [Question GWE3 on the GA questionnaire asks for the same information.]

The question does not provide a definition of what constitutes *accurate* weather information. The perception of the extent to which the weather information was accurate or sufficiently accurate for their needs is likely to vary among respondents. As noted before, cognitive testing of the question would have been appropriate.

G

Air Carrier Questionnaire

The following excerpt from the *NAOMS Reference Report* (Battelle, 2007) introduces the air carrier questionnaire, which is reprinted here.

> Interviewing of the air carrier (AC) pilots began in March 2001. The initial interview sample was split between pilots who were randomly selected for one interview and pilots who were asked to complete the interview once every three months. This appendix includes a copy of the AC questionnaire. The questionnaire consisted of four sections that corresponded with general topics covered in the general aviation (GA) questionnaire: Section A addressed pilot qualifications and experience; Section B addressed safety events; Section C addressed a specific focus topic;[1] and Section D offered pilots an opportunity to provide feedback on the interview process and the questionnaire. This appendix contains a copy of the air carrier questionnaire.

[1] Two Section Cs were prepared during the course of this survey: one pertains to In-close Approach Changes (ICACs), and the other pertains to the development of baseline safety for the CAST-JIMDAT.

Air Carrier Questionnaire

Section A: Background Questions

TIME BEGUN ....(MILITARY) .... __ __ : __ __
(FILLS)

INTERVIEWER: DATE OF INTERVIEW IS BEING RECORDED AS (START DATE).
IS THIS THE CORRECT DATE?

YES.. 1
NO .... (RECORD DATE OF INTERVIEW) 0

START DATE .... __ __ / __ __ / __ __
MONTH DAY YEAR

.... S
TART DATE = 30/90 DAYS BEFORE END DATE

END DATF .... __ __ / __ __ / __ __
(FILLS) MONTH DAY YEAR
END DATE = DAY
BEFORE DAY OF INTERVIEW

SECTION A: BACKGROUND QUESTIONS

INTRODUCTION:

For this survey most of the questions will refer to (30/90) days prior to today. Therefore, whenever I say the "last (TIME PERIOD), I am referring to the period from (START DATE) through (END DATE).

I am now going to ask you a few questions about the commercial flying that you did during the last (TIME PERIOD).

A1.	During the last (TIME PERIOD), how many **hours** did you fly as a crewmember on commercial aircraft?	# HOURS IN TIME PERIOD.............. __ __ __
	PROMPT IF 30 DAYS>100, 90 DAYS>300: *I'd just like to verify. You said you flew (HOURS A1) hours during the last (TIME PERIOD) as a crewmember on a commercial aircraft. Is this correct?*	*NO.......... 0* *YES.......... (A2).......... 1* *RF.......... (A2).......... 7* *DK.......... (A2).......... 8*
A1 *NEW*	*During the last (TIME PERIOD), how many hours did you fly as a crewmember on a commercial aircraft?*	*#HOURS.......... __ __ __* *RF.......... 997* *DK.......... 998*

A2. During the last (TIME PERIOD), how many **legs** did you fly as a crewmember on commercial aircraft?

LEGS IN TIME PERIOD........ |__|__|__|

A2.1 During the last (TIME PERIOD), how many of the (#A2) legs you flew involved taking off or landing at an airport outside the United States?

NOTE: THE UNITED STATES MEANS THE 50 STATES AND WASHINGTON DC, BUT DOES NOT INCLUDE US TERRITORIES.

LEGS OUTSIDE U.S........ |__|__|__|

NUMBER OF LEGS IN A2.1 MUST BE LESS THAN OR EQUAL TO LEGS IN A2.

A3. Please tell me the makes, models and series for all of the aircraft you flew commercially as a crewmember during the last (TIME PERIOD)? RECORD VERBATIM IN COLUMN A, THEN ASK PROMPT.

PROMPT A3_A1: Did you fly any other makes, models or series of aircraft commercially during the last (TIME PERIOD)?	*YES........ (ASK PROMPT A3_A2)........ 1* *NO........ (ASK B)........ 0* *RF........ 7* *DK........ 8*

PROMPT A3_A2: Please tell me the next aircraft make, model and series you flew commercially as a crewmember during the last (TIME PERIOD)? RECORD IN COLUMN A

A. MAKE/MODEL/SERIES (NOTE; MAKE/MODEL/SERIES DROP DOWN SCREEN INCREASED WITH THIS VERSION)	B. During the last (TIME PERIOD), what **percent** of the (HRS IN A1) did you fly the (MAKE/ MODEL/SERIES)?
1st ______	|__|__|__| %
2nd ______	|__|__|__| %
3rd ______	|__|__|__| %
4th ______	|__|__|__| %
5th ______	|__|__|__| %
6th ______	|__|__|__| %
	THE TOTAL PERCENT OF A3-B SHOULD BE 100.

INTRODUCTION:
During the last (TIME PERIOD), you may have transported passengers or cargo, or conducted other flight operations. We would like to understand what types of operations you flew.

A4. During the last (TIME PERIOD), what percent of the (HRS IN A1) did you fly as a crewmember on flights with revenue passengers?

% WITH REVENUE PASSENGERS .... |__|__|__|

A5. During the last (TIME PERIOD), what percent of the (HRS IN A1) did you work as a crewmember on flights that carried only cargo or freight and did **not** carry revenue passengers?

% CARGO/FREIGHT W/O PASSENGERS .... |__|__|__|

A6. During the last (TIME PERIOD), what percent of the (HRS IN A1) did you work as a crewmember on flights that carried no revenue passengers or cargo, such as maintenance flights, ferry flights, or repositioning flights?

% NO PASSENGER OR CARGO .... |__|__|__|

THE TOTAL PERCENT OF A4, A5, AND A6 SHOULD BE 100.

A. What type of flights were these?

SPECIFY: ______________________________

A7. During the last (TIME PERIOD), did you fly a commercial aircraft (READ QUESTIONS)?

	YES	NO	RF	DK
a. as a captain	1	0	7	8
b. as a first officer	1	0	7	8
c. as a flight engineer or second officer	1	0	7	8
d. as a relief pilot	1	0	7	8
e. in any other capacity (SPECIFY)	1	0	7	8

1. What was that capacity?

A7a THROUGH A7e CANNOT ALL BE ANSWERED NO.

SPECIFY: ______________________________

INTERVIEWER: CAN INCLUDE CHECK PILOT.

A7.1 Which of the following three categories best describes the number of airplanes currently operated by your airline? Please do not include airplanes operated by code-share partners. READ CATEGORIES.

NOTE: WE ARE ONLY INTERESTED IN AIRPLANES CURRENTLY BEING USED, NOT THOSE IN STORAGE.

PROBE IF PILOT FLEW FOR MORE THAN ONE AIRLINE IN TIME PERIOD: Please tell me the number of airplanes currently operated by the airline **that you flew the most hours for** in the last (TIME PERIOD).

350 airplanes or more...1
150 to 349 airplanes..2
149 or less airplanes 3
RF..7
DK..8

A8. Approximately how many **hours** in total have you flown a **commercial aircraft during your career**?

TOTAL HOURS DURING CAREER................ |_|_|_|_|_|

Air Carrier Questionnaire

Section B:
Safety Related Events

SECTION B: SAFETY RELATED EVENTS

INTRODUCTION:

My next questions are about safety related events. In answering these questions, please report only events that **you experienced on a commercial aircraft on which you were a crewmember**. The first of these questions are about **equipment-related events.**

ER1. How many times during the last (TIME PERIOD) did an aircraft on which you were a crewmember divert to an alternate airport or return to land because of an aircraft equipment problem?

EQUIPMENT PROBLEMS.. |__|__|__|

A. What systems caused the diversion or return to land?

SPECIFY: ____________________

ER2. How many times during the last (TIME PERIOD) did an aircraft on which you were a crewmember experience a spill, fire, fumes, or aircraft damage due to transporting hazardous materials?

HAZMAT .... |__|__|__|

IF 0, SKIP TO ER3.

A. (How many of these [# in ER2] times were the spills, fire, fumes or aircraft damage/Was this spill, fire, fumes or aircraft damage) in the cargo compartment?

IN CARGO COMPARTMENT.................................. |__|__|__|

THE AMOUNT IN ER2A CANNOT BE GREATER THAN THE AMOUNT IN ER2.

B. (How many of these [# in ER2] times were spills, fire, fumes or aircraft damage/Was this spill, fire, fumes or aircraft damage) in the passenger compartment?

IN PASSENGER COMPARTMENT.......................... |__|__|__|

THE AMOUNT IN ER2A AND ER2B COMBINED CANNOT BE GREATER THAN THE AMOUNT IN ER2.

C. (How many of these [# IN ER2] times were the spills, fire, fumes or aircraft damage/Was the spill, fire, fumes or aircraft damage) caused because the hazardous materials in question were out of compliance with regulations?

OUT OF COMPLIANCE WITH REGULATIONS ...... |__|__|__|

THE AMOUNT IN ER2C CANNOT BE GREATER THAN THE AMOUNT IN ER2.

ER3. How many times during the last (TIME PERIOD) did an aircraft on which you were a crewmember experience a cargo shift

CARGO SHIIFTS.. |__|__|__|

ER4. How many times during the last (TIME PERIOD) did an in-flight aircraft on which you were a crewmember experience uncommanded movements of any of the following devices (READ QUESTIONS)?

a. Uncommanded movements of the elevators? .... # ELEVATORS .... |__|__|__|

b. Uncommanded movements of the rudder? .... # RUDDER .... |__|__|__|

c. Uncommanded movements of the ailerons? .... # AILERONS .... |__|__|__|

d. Uncommanded movements of the spoilers? .... # SPOILERS .... |__|__|__|

e. Uncommanded movements of the speedbrakes? . # SPEEDBRAKERS .... |__|__|__|

f. Uncommanded movements of the trim tabs? .... # TRIM TABS .... |__|__|__|

g. Uncommanded movements of the flaps? .... # FLAPS .... |__|__|__|

h. Uncommanded movements of the slats? .... # SLATS .... |__|__|__|

i. Did any other devices have uncommanded movements during the last (TIME PERIOD)?

YES 1
NO .... (SKIP TO ER5) 0
RF .... (SKIP TO ER5) 7
DK .... (SKIP TO ER5) 8

1. Which devices?

SPECIFY: ____________________

2. **FOR EACH DEVICE LISTED IN ER4i1:** How many times did (DEVICE LISTED IN ER4i1) perform uncommanded movements during the last (TIME PERIOD)?

UNCOMMANDED MOVEMENTS .... |__|__|__|

ER5. How many times during the last (TIME PERIOD) did an inflight aircraft on which you were a crewmember experience smoke, fire, or fumes that originated in any of the following areas (READ QUESTIONS):

A. the engine or nacelle? .... # IN ENGINE OR NACELLE .... |__|__|__|

IF 0, SKIP TO ER5B.

1. (Of the [# in ER5A] times there was smoke, fire, or fumes in the engine or nacelle, how many involved/Did the smoke, fire, or fumes in the engine or nacelle involve) electrical components or wiring?

SMOKE/FIRE/FUMES .... |__|__|__|

THE AMOUNT IN ER5A1 CANNOT BE GREATER THAN THE AMOUNT IN ER5A.

B. the flight deck? .... # IN FLIGHT DECK .... |__|__|__|

IF 0, SKIP TO ER5C.

1. (Of the [# in ER5B] times there was smoke, fire, or fumes in the flight deck, how many involved/Did the smoke, fire, or fumes in the flight deck involve) electrical components or wiring?

SMOKE/FIRE/FUMES .... |__|__|__|

THE AMOUNT IN ER5B1 CANNOT BE GREATER THAN THE AMOUNT IN ER5B.

C. the cargo hold? .... # IN CARGO HOLD .... |__|__|__|
IF 0, SKIP TO ER5D.

1. (Of the [# in ER5C] times there was smoke, fire, or fumes in the cargo hold, how many involved/Did the smoke, fire, or fumes in the cargo hold involve) electrical components or wiring? SMOKE/FIRE/FUMES .... |__|__|__|

THE AMOUNT IN ER5C1 CANNOT BE GREATER THAN THE AMOUNT IN ER5C.

D. the galley? .... # IN GALLEY .... |__|__|__|
IF 0, SKIP TO ER5E.

1. (Of the [# in ER5D] times there was smoke, fire, or fumes in the galley, how many involved/Did the smoke, fire, or fumes in the galley involve) electrical components or wiring? SMOKE/FIRE/FUMES .... |__|__|__|

THE AMOUNT IN ER5D1 CANNOT BE GREATER THAN THE AMOUNT IN ER5D.

E. elsewhere in the passenger compartment? ...... # IN ELECTRICAL COMPONENETS OR WIRING ..... |__|__|__|
IF 0, SKIP TO ER5F.

1. (Of the [# in ER5E] times there was smoke, fire, or fumes elsewhere in the passenger compartment, how many involved/Did the smoke, fire, or fumes elsewhere in the passenger compartment involve) electrical components or wiring? SMOKE/FIRE/FUMES .... |__|__|__|

THE AMOUNT IN ER5E1 CANNOT BE GREATER THAN THE AMOUNT IN ER5E.

F. During the last (TIME PERIOD), how many times did an inflight aircraft on which you were a crewmember experience smoke, fire or fumes that originated other than in the engine or nacelle, flight deck, cargo hold, galley, or passenger compartment? # ORIGINATE OTHER PLACES .... |__|__|__|

1. Where did the smoke, fire or fumes originate? SPECIFY.

SPECIFY: ______________________________

ER6. During the last (TIME PERIOD), how many times did an inflight aircraft on which you were a crewmember experience a precautionary engine shutdown? # PRECAUTIONARY ENGINE SHUTDOWNS .... |__|__|__|

ER7. During the last (TIME PERIOD) how many times did an inflight aircraft on which you were a crewmember experience a total engine failure? # TOTAL ENGINE FAILURE .... |__|__|__|

INTRODUCTION:

The following questions relate to **turbulence**.

During the last (TIME PERIOD), how many times did an aircraft on which you were a crewmember (READ QUESTION)?

TU1. Encounter severe turbulence that caused large abrupt changes in altitude, airspeed, or attitude .. # CAUSED ABRUPT CHANGES.................................. |__|__|__|

IF 0, SKIP TO TU2.

A. (Of the [#in TU1] severe turbulence encounters, how many occurred/Did this severe turbulence encounter occur) in I.M.C. conditions? I.M.C. = INSTRUMENT METEOROLOGICAL CONDITIONS

IN IMC CONDITIONS .... |__|__|__|

THE AMOUNT IN TU1A CANNOT BE GREATER THAN THE AMOUNT IN TU1.

B. (Of the [# in TU1] severe turbulence encounters, how many occurred/Did this severe turbulence encounter occur) in clear air?

IN CLEAR AIR .... |__|__|__|

THE AMOUNT IN TU1A AND TU1B CANNOT BE GREATER THAN THE AMOUNT IN TU1.

TU2. Encounter wake turbulence that resulted in 10 or more degrees of aircraft roll .... # RESULTING IN AIRCRAFT ROLL .... |__|__|__|

INTRODUCTION:

The next few questions are about **weather-related events while airborne**.

During the last (TIME PERIOD), how many times did an aircraft on which you were a crewmember (READ QUESTION)?

WE1. Lack accurate weather information when crewmembers needed it while airborne .... # LACK WEATHER INFORMATION .... |__|__|__|

IF 0, SKIP TO WE2.

A. (Of the [# WE1] times when crewmembers lacked accurate weather information while airborne, how many involved non-U.S. airports or controllers?/ Did this time when crewmembers lacked accurate weather information while airborne involve a non-U.S. airport or controller?)

INVOLVE NON-US AIRPORT OR CONTROLLER ... |__|__|__|

THE AMOUNT IN WE1A CANNOT BE GREATER THAN THE AMOUNT IN WE1.

B. (Of the [# WE1] times when crewmembers lacked accurate weather information while airborne, how many involved ATIS?/Did this time when crewmembers lacked accurate weather information while airborne involve ATIS?) # INVOLVE ATIS.......... |_|_|_|

THE AMOUNT IN WE1A AND WE1B COMBINED CANNOT BE GREATER THAN THE AMOUNT IN WE1.

WE2. Fail to receive A.T.C. approval for a request to avoid severe weather.......... # FAIL RECEIVE ATC APPROVAL.......... |_|_|_|

IF 0, SKIP TO WE3.

A. (Of the [# WE2] times crewmembers failed to receive A.T.C. approval to avoid severe weather, how many times was emergency authority invoked in these situations/Was emergency authority invoked in this situation? # EMERGENCY AUTHORITY INVOKED.......... |_|_|_|

THE AMOUNT IN WE2A CANNOT BE GREATER THAN THE AMOUNT IN WE2.

WE3. Divert to an alternate airfield because of weather .... # DIVERT TO ALTERNATE AIRFIELD.......... |_|_|_|

WE4. Experience airframe icing that reduced the aircraft's ability to maintain altitude, speed, stability, or directional control.......... # EXPERIENCE AIRFRAME ICING.......... |_|_|_|

WE5. Encounter windshear or a microburst condition that resulted in an airspeed deviation of 15 knots or greater.......... # ENCOUNTER WINDSHEAR/MICROBURST.......... |_|_|_|

WE6. Encounter windshear or a microburst condition that resulted in a windshear avoidance maneuver .... # RESULT IN WINDSHEAR AVOIDANCE.......... |_|_|_|

IF A4=0, SKIP TO AC1.

INTRODUCTION:
The next few questions are about **passenger-related events**.

During the last (TIME PERIOD), how many times did an **in-flight** aircraft on which you were a crewmember (READ QUESTIONS):

CP1. Expedite landing or divert to an alternate airport due to a passenger medical emergency.......... # DUE TO PASSENGER MEDICAL EMERGENCY.......... |_|_|_|

CP2. Expedite landing or divert to an alternate airport due to a passenger disturbance.......... # DUE TO PASSENGER DISTURBANCE.......... |_|_|_|

CP3. During the last (TIME PERIOD), how many times did a crewmember leave the cockpit to handle a passenger disturbance on an inflight aircraft on which you were a crewmember .... # CREWMEMBERS LEAVE COCKPIT.......... |_|_|_|

INTRODUCTION:

The next few questions are about **airborne conflicts**.

During the last (TIME PERIOD), how many times did an aircraft on which you were a crewmember (READ QUESTION)?

AC1.	Experience a bird strike	# BIRD STRIKES ⎿⏊⏊⏌
AC2.	Perform an evasive action to avoid an imminent in-flight collision with another aircraft that was never closer than 500 feet including evasive action in response to a TCAS advisory?	# EVASIVE ACTIONS ⎿⏊⏊⏌
AC3.	Experience less than 500 feet of separation from another aircraft while both aircraft were airborne	# LESS THAN 500 FEET SEPARATION ⎿⏊⏊⏌

INTRODUCTION:

The next few questions are about **ground operations**.

During the last (TIME PERIOD), how many times did an aircraft on which you were a crewmember (READ QUESTION)?

GE1.	Go off the edge of a runway or taxiway while taxiing	# GO OFF EDGE RUNWAY/TAXIWAY ⎿⏊⏊⏌
GE2.	Collide or nearly collide with a ground vehicle?	# COLLIDE WITH GROUND VEHICLE ⎿⏊⏊⏌ **IF 0, SKIP TO GE3.**
	A. (Of the [# in GE2] near collisions with a ground vehicle, how many occurred/Did this near collision with a ground vehicle occur) while your aircraft was on the ramp, apron or in the gate area?	# ON RAMP/APRON/GATE AREA ⎿⏊⏊⏌ ***THE AMOUNT IN GE2A CANNOT BE GREATER THAN THE AMOUNT IN GE2.***
	B. (Of the [# in GE2] near collisions with a ground vehicle, how many occurred/Did this near collision with a ground vehicle occur) while your aircraft was on the taxiway?	# ON TAXIWAY ⎿⏊⏊⏌ ***THE AMOUNT IN GE2A AND GE2B COMBINED CANNOT BE GREATER THAN THE AMOUNT IN GE2.***
	C. (Of the [# in GE2] near collisions with a ground vehicle, how many occurred/Did this near collision with a ground vehicle occur) while your aircraft was on the runway?	# ON RUNWAY ⎿⏊⏊⏌ ***THE AMOUNT IN GE2A, GE2B, AND GE2C COMBINED CANNOT BE GREATER THAN THE AMOUNT IN GE2.***
GE3.	Skid, slide, or hydroplane resulting in a significant increase in stopping distance during landing	# SKID/SLIDE/HYDROPLANE ⎿⏊⏊⏌
GE4.	Experience a rejected takeoff	# REJECTED TAKEOFFS ⎿⏊⏊⏌

GE5. Go off the edge of a runway while taking off or landing .... # GO OFF EDGE OF RUNWAY .... |__|__|__|

GE6. Go off the end of the runway .... # GO OFF END OF RUNWAY .... |__|__|__|

GE7. Inadvertently enter an active runway .... # ENTER ACTIVE RUNWAY .... |__|__|__|

GE8. Begin takeoff roll while another aircraft occupied or was crossing the same runway .... # TAKEOFF ROLL WITH OCCUPIED RUNWAY .... |__|__|__|

GE9. Land while another aircraft occupied or was crossing the same runway .... # LAND ON OCCUPIED RUNWAY .... |__|__|__|

GE10. Nearly experience a ground collision with another aircraft while both aircraft were on the ground .... # NEAR GROUND COLLISION .... |__|__|__|

IF 0, SKIP TO AH1.

A. (Of the [# in GE10] near collisions with another aircraft, how many occurred/Did this near collision with another aircraft occur) while your aircraft was on the ramp, apron or in the gate area? # ON RAMP/APRON/GATE AREA .... |__|__|__|

THE AMOUNT IN GE10A CANNOT BE GREATER THAN THE AMOUNT IN GE10.

B. (Of the [# in GE10] near collisions with another aircraft, how many occurred/Did this near collision with another aircraft occur) while your aircraft was on the taxiway? # ON TAXIWAY .... |__|__|__|

THE AMOUNT IN GE10A AND GE10B COMBINED CANNOT BE GREATER THAN THE AMOUNT IN GE10.

C. (Of the [# in GE10] near collisions with another aircraft, how many occurred/Did this near collision with another aircraft occur) while your aircraft was on the runway? # ON RUNWAY .... |__|__|__|

THE AMOUNT IN GE10A, GE10B, AND GE10C COMBINED CANNOT BE GREATER THAN THE AMOUNT IN GE10.

INTRODUCTION:

The next few questions are about **aircraft handling-related events.**

During the last (TIME PERIOD), how many times did an aircraft on which you were a crewmember (READ QUESTION)?

AH1. Use some of its reserve fuel as defined by the F.A.Rs .... # USE RESERVE FUEL .... |__|__|__|

AH2. Accept an A.T.C. clearance that the aircraft could not comply with because of its performance limits .... # ACCEPT CLEARANCE NOT COMPLY WITH .... |__|__|__|

AH3. Lose sight of another aircraft from which the aircrew was trying to maintain visual separation # LOSE SIGHT OF AIRCRAFT .... |__|__|__|

IF 0, SKIP TO AH4.

A. (Of the [# in AH3] times an aircraft lost sight of another aircraft, how many occurred/Did losing sight of another aircraft occur) in marginal visual conditions of 3 miles or less? # IN MARGINAL VISUAL CONDITONS .... |__|__|__|

THE AMOUNT IN AH3A CANNOT BE GREATER THAN THE AMOUNT IN AH3.

AH4.	Inadvertently land without clearance at an airport with an active control tower	# LAND W/O CLEARANCE .... |__|__|__|
AH5.	Inadvertently begin takeoff roll without A.T.C. clearance at an airport with an active control tower	# TAKEOFF ROLL W/O CLEARANCE .... |__|__|__|
AH6	Inadvertently deviate from an assigned routing or A.T.C. vector for one minute or more	# DEVIATIONS .... |__|__|__|
AH7.	Experience a tail strike on landing	# TAIL STRIKES ON LANDING .... |__|__|__|
AH8.	Experience a tail strike on takeoff	# TAIL STRIKES ON TAKEOFF .... |__|__|__|
AH9.	Experience a hard landing	# HARD LANDINGS .... |__|__|__|
AH10.	Take off with an out-of-limit center of gravity	# TAKE-OFF OUT-OF-LIMIT CENTER OF GRAVITY |__|__|__|
AH11.	Take-off overweight	# TAKE-OFF OVERWEIGHT .... |__|__|__|
AH12.	Commence take-off roll with an improper aircraft configuration	# WITH IMPROPER CONFIGURATION .... |__|__|__|
AH13.	Experience an unusual attitude for any reason	# UNUSUAL ATTITUDE .... |__|__|__|
AH14.	Experience a valid stall warning or stick shaker activation	# STALL WARNING/STICK SHAKER ACTIVATION ... |__|__|__|
AH15.	Nearly collide with terrain or a ground obstruction while airborne?	# NEAR COLLISIONS/GROUND .... |__|__|__| **IF 0, SKIP TO AD1.**

INTERVIEWER: INCLUDES BUILDINGS

A. (Of the [# in AH15] near collisions with terrain or a ground obstruction, how many were/Was this near collision with terrain or a ground obstruction)-brought to your attention by A.T.C.?

ATC BROUGHT TO YOUR ATTENTION .... |__|__|

THE AMOUNT IN AH15A CANNOT BE GREATER THAN THE AMOUNT IN AH15.

B. (Of the [# in AH15] near collisions with terrain or a ground obstruction, how many were/Was this near collision with terrain or a ground obstruction) detected through direct sighting of the ground or obstruction?

DETECTED THROUGH DIRECT SIGHTING .... |__|__|

THE AMOUNT IN AH15A AND AH15B COMBINED CANNOT BE GREATER THAN THE AMOUNT IN AH15.

C. (Of the [# in AH15] near collisions with terrain or a ground obstruction, how many were/Was this near collision with terrain or a ground obstruction)-detected through activation of G.P.W.S. or E.G.P.W.S.? ....

DETECTED THROUGH GPWS/EGPWS .... |__|__|

THE AMOUNT IN AH15A, AH15B, AND AH15C COMBINED CANNOT BE GREATER THAN THE AMOUNT IN GE10.

1. (How many of these [# in AH15c] near collisions were/Was this near collision) detected through activation of E.G.P.W.S.?

DETECTED THROUGH ACTIVATION OF EGPWS |__|__|__|

THE AMOUNT IN AH15C1 CANNOT BE GREATER THAN THE AMOUNT IN AH15C.

INTRODUCTION:
The next few questions are about **altitude deviations**.

How many times during the last (TIME PERIOD) did an aircraft on which you were a crewmember (READ QUESTIONS)?

AD1. Inadvertently deviate from an assigned altitude by more than 300 feet?.. # ALTITUDE DEVIATIONS .... |__|__|__|
IF 0, SKIP TO AD2.

A. (Of the [# in AD1] deviations from an assigned altitude, how many were/Was this deviation from an assigned altitude) in response to a TCAS Resolution Advisory? # IN RESPONSE TO TCAS.. |__|__|__|

THE AMOUNT IN AD1A CANNOT BE GREATER THAN THE AMOUNT IN AD1.

AD2. Descend below Minimum Safe Altitude when you were **not** following A.T.C. radar vectors........ # NOT FOLLOWING ATC RADAR VECTORS .... |__|__|__|

INTRODUCTION:
The next few questions are about **interactions with air traffic control**.

AT1. During the last (TIME PERIOD), how many times was an aircraft on which you were a crewmember unable to communicate with A.T.C. in a time-critical situation because of frequency congestion? # UNABLE TO COMMUNICATE WITH ATC .... |__|__|__|
IF 0, SKIP TO AT2.

These problems may have occurred on the ground, or while airborne in the terminal area, or while en route. I'm going to ask you about each.

A. (Of these [# in AT1] times you were unable to communicate with A.T.C. in a time-critical situation because of frequency congestion, how many occurred/Did the time you were unable to communicate with A. T.C in a time critical situation because of frequency congestion occur) **while on the ground**? # WHILE ON GROUND.. |__|__|__| # TIMES

THE AMOUNT IN AT1A CANNOT BE GREATER THAN THE AMOUNT IN AT1.

B. (Of these [# in ATI1] times you were unable to communicate with A.T.C. in a time-critical situation because of frequency congestion, how many occurred/Did the time you were unable to communicate with A. T.C in a time critical situation because of frequency congestion occur) **while airborne in the terminal area**? .... # WHILE AIRBORNE .... |__|__|__| # TIMES

THE COMBINED TOTALS IN AT1A AND AT1B CANNOT BE GREATER THAN 100.

C. (Of these [# in ATI1] times you were unable to communicate with A.T.C. in a time-critical situation because of frequency congestion, how many occurred/Did the time you were unable to communicate with A. T.C in a time critical situation because of frequency congestion occur) **while en route**? .... # WHILE EN ROUTE .... |__|__|__| # TIMES

THE COMBINED TOTALS IN AT1A, AT1B, AND AT1C CANNOT BE GREATER THAN 100.

AT2. How many times during the last (TIME PERIOD) did an aircraft on which you were a crewmember fly at an undesirably high altitude or airspeed on approach due to an A.T.C. clearance .... # HIGH ALTITUDE OR AIRSPEED .... |__|__|__|

NOTE TO INTERVIEWERS: THIS INCLUDES BUT MAY NOT BE LIMITED TO "SLAM DUNK" APPROACHES.

Air Carrier Questionnaire

Section C:
In-close Approach Changes

SECTION C: IN-CLOSE APPROACH CHANGES

INTRODUCTION:

My next questions are about **clearance changes received on approach** within 10 miles of the runway threshold **that the flight crew did not request**.

IC1. During the last (TIME PERIOD), how many times did an aircraft on which you were a crewmember receive an unrequested clearance change to runway assignment, altitude restrictions or airspeed within 10 miles of the runway threshold?

UNREQUESTED CLEARANCE CHANGES ....|__|__|__|

IF 00, DK OR RF, SKIP TO SECTION D.
IF 01, CONTINUE WITH ROUTE A.
IF 02 OR MORE, SKIP TO ROUTE B.

ROUTE A—ONLY ONE CHANGE

A. Was this unrequested clearance change declined?

YES..........................(SKIP TO SECTION D)..........................001
NO..000
RF..........................(SKIP TO SECTION D)..........................997
DK..........................(SKIP TO SECTION D)..........................998

B. Did this unrequested clearance change result in (READ QUESTIONS)?

	YES	NO	RF	DK
1. An unstabilized approach.........................	1	0	7	8
2. A go-around or missed approach............	1	0	7	8
3. An airborne conflict..................................	1	0	7	8
4. A wake turbulence encounter...................	1	0	7	8
5. Landing with out-of-limit tailwinds or crosswinds..	1	0	7	8
6. Landing on a wrong runway.....................	1	0	7	8
7. Landing long or fast..................................	1	0	7	8
8. Landing without clearance	1	0	7	8
9. A conflict on the ground with another aircraft or ground vehicle?........................	1	0	7	8
10. Any other undesirable event after the clearance change?..................................	1 ASK a.	0 SKIP TO IC2.	7 SKIP TO IC2.	8 SKIP TO IC2.

a. What events occurred?

SPECIFY: ____________________

SKIP TO IC2.

ROUTE B—TWO OR MORE CHANGES

A. Of the (# IN IC1) unrequested clearance changes, how many, if any, were declined?

UNREQUESTED CLEARANCE CHANGES .... |__|__|

IF NUMBER IN IC1A=NUMBER IN IC1, DK or RF, SKIP TO SECTION D.

IF ONLY ONE CHANGE REMAINS, GO TO ROUTE A, IC1B.

THE NUMBER OF UNREQUESTED CLEARANCE CHANGES WAS (NUMBER IC1) SO THE NUMBER OF UNREQUESTED CLEARANCE CHANGES THAT WERE DECLINED HAS TO BE (NUMBER IN IC1) OR FEWER.

B. How many of the accepted clearance changes resulted in (READ QUESTIONS)? IF 01 OR GREATER, ASK C.

THE ANSWERS IN IC1B 1-10 CANNOT BE GREATER THAN IC1 MINUS IC1A.

C. Did (this/any of these) (EVENT) happen in the most recent **accepted** clearance change?

	# CHANGES	YES	NO	RF	DK
1. An unstabilized approach	|__|__|	1	0	7	8
2. A go-around or missed approach	|__|__|	1	0	7	8
3. An airborne conflict	|__|__|	1	0	7	8
4. A wake turbulence encounter	|__|__|	1	0	7	8
5. Landing with out-of-limit tailwinds or crosswinds	|__|__|	1	0	7	8
6. Landing on a wrong runway	|__|__|	1	0	7	8
7. Landing long or fast	|__|__|	1	0	7	8
8. Landing without clearance	|__|__|	1	0	7	8
9. A conflict on the ground with another aircraft or ground vehicle?	|__|__|	1	0	7	8
10. Any other undesirable event after the clearance change?	|__|__|	1 ASK a.	0 SKIP TO IC2.	7 SKIP TO IC2.	8 SKIP TO IC2.

IF NONE, SKIP TO IC2.IF ≥1, ASK a.

a. What events occurred?

SPECIFY: ______________________________

INTRODUCTION:

(My next questions are about **this accepted clearance** change that we have been talking about./My next questions are about the **most recent clearance change** that the flight crew **accepted.**)

IC2. At which airport did this event occur? NAME OF AIRPORT: ______________

A. Please tell me the location identifier for (AIRPORT). AIRPORT LOCATION ID: ______________

IC3. **ASK ONLY IF TWO OR MORE MODELS REPORTED IN A3. IF ONLY ONE MODEL, SKIP TO IC4.**

Which model aircraft were you flying when this event occurred, the (LIST MODELS IN A3A)? NAME/MODEL: ______________
CODE MODEL FROM A3A

IC4. Were you a crewmember on an F.M.S. or F.M.C. equipped aircraft at the time of this event?

YES........1
NO........(SKIP TO IC8)........0
RF........(SKIP TO IC8)........7
DK........(SKIP TO IC8)........8

A. Was the F.M.S. or F.M.C. that was being used capable of storing multiple routes?

YES........1
NO........(SKIP TO IC8)........0
RF........(SKIP TO IC8)........7
DK........(SKIP TO IC8)........8

B. Are the navigation and communication frequency changes in this aircraft made through the F.M.S. or F.M.C.?

YES........1
NO........0
RF........7
DK........8

IC5. In response to this clearance change, did the flightcrew reprogram or attempt to reprogram the F.M.S. or F.M.C.

YES........1
NO........(SKIP TO IC8)........0
RF........(SKIP TO IC8)........7
DK........(SKIP TO IC8)........8

IC6. When programming changes were made or attempted, (READ QUESTIONS)?

	YES	NO	RF	DK
A. Did the inputs load properly........	1	0	7	8
B. Was it possible to complete the programming within available time........	1	0	7	8
C. Were all of the programming inputs cross-checked by other crewmembers?........	1	0	7	8
D. Were there other programming difficulties........	1	0	7	8
	ASK 1.	SKIP TO IC7.		

1. Please describe these difficulties.

SPECIFY: ______________

IC7. Overall, did the F.M.S. or F.M.C. assist you in complying with the clearance change?

YES 1
NO 0
RF 7
DK 8

ONLY IF ROUTE B IC1A IS 2 OR GREATER, READ INTRODUCTION:

INTRODUCTION:

Before we continue, I want to remind you that these questions are still about the **most recent** unrequested clearance change within 10 miles of the runway threshold.

IC8. Was the aircraft on an instrument approach prior to the clearance change?

YES 1
NO .... (SKIP TO IC9) 0
RF .... (SKIP TO IC9) 7
DK .... (SKIP TO IC9) 8

A. Did this change involve a change from an instrument approach to a visual approach?

YES 1
NO .... (SKIP TO IC10) 0
RF .... (SKIP TO IC10) 7
DK .... (SKIP TO IC10) 8

IC9. Did this change involve a change from a visual approach to an instrument approach?

YES 1
NO 0
RF 7
DK 8

IC10. Was the aircraft programmed for an auto-coupled approach at the time of the clearance change?

YES 1
NO 0
RF 7
DK 8
NA 9

IC11. Did this clearance change the aircraft's runway assignment?

YES 1
NO .... (SKIP TO IC12) 0
RF .... (SKIP TO IC12) 7
DK .... (SKIP TO IC12) 8

A. Did the runway reassignment involve a change from one runway to another parallel runway

YES 1
NO 0
RF 7
DK 8

IC12. Did this clearance change the aircraft's altitude assignment?

YES 1
NO 0
RF 7
DK 8

IC13. Did this clearance change the aircraft's airspeed assignment?

YES 1
NO 0
RF 7
DK 8

ONLY IF ROUTE B IC1A IS 2 OR GREATER, READ INTRODUCTION:

INTRODUCTION:

Once again, before we continue, I want to remind you that these questions are still about the **most recent** unrequested clearance change within 10 miles of the runway threshold.

IC14. In response to this clearance change, did the flightcrew (READ QUESTIONS)?

	YES	NO	RF	DK
A. Change a navigational aid frequency	1 (ASK 1)	0 (SKIP TO B)	7 (SKIP TO B)	8 (SKIP TO B)
1. Confirm the identity of the new navaid	1	0	7	8
B. Change the A.T.C. communication frequency	1	0	7	8
C. Revise the approach briefing	1	0	7	8
D. Change the airplane configuration	1	0	7	8
E. Disconnect any of the automated control systems?	1	0	7	8

IC15. Was the flight crew given a reason for the clearance change?

YES 1
NO .... (SKIP TO IC16) 0
RF .... (SKIP TO IC16) 7
DK .... (SKIP TO IC16) 8

A. Was one of the reasons given (READ QUESTIONS)?

	YES	NO	RF	DK
1. Wake turbulence avoidance	1	0	7	8
2. Maintaining traffic flow and separation	1	0	7	8
3. Providing a runway favorable to your gates	1	0	7	8
4. A change in active runways	1	0	7	8
5. Weather or wind factors	1	0	7	8
6. Noise abatement factors	1	0	7	8
7. A.T.C. equipment problems	1	0	7	8
8. Was any other reason given for the clearance change	1 ASK a	0 SKIP TO IC16	7 SKIP TO IC16	8 SKIP TO IC16

a. What reasons were given?

SPECIFY:______________________________

IC16. Did responding to the clearance change (READ QUESTIONS)?

	YES	NO	RF	DK
A. reduce the quality of cockpit coordination.........	1	0	7	8
B. reduce situational awareness	1	0	7	8
C. Compromise traffic watch	1	0	7	8
D. Was safety compromised in any other way.	1 ASK 1.	0	7	8

(NO, RF, DK: SKIP TO SECTION D.)

1. How was safety compromised?

SPECIFY:____________________________________

Air Carrier Questionnaire

Section C:
JIMDAT Questions

SECTION C: JIMDAT QUESTIONS

INTRODUCTION:

In the next section, I will be asking you some questions about your flying experience and training as it relates to terminal operations and instrument approaches. As we go forward, please limit you answers to those things that you personally experienced.

JD1. Is the aircraft you flew (most) during the last 60 days equipped with G.P.W.S?

GPWS = ground proximity warning system

NO .... (SKIP TO JD2) 0
YES 1
RF .... (SKIP TO JD2) 7
DK .... (SKIP TO JD2) 8

A. Is it equipped with a terrain display, such as you find in an enhanced G.P.W.S, or Terrain Avoidance Warning System, also known as TAWS (taws)?

NO .... (SKIP TO JD2) 0
YES 1
RF .... (SKIP TO JD2) 7
DK .... (SKIP TO JD2) 8

B. Does your airline require the terrain display to be selected during takeoff at specific airports?

NO OR NEVER .... (SKIP TO JD2) 0
YES OR SOMETIMES 1
RF .... (SKIP TO JD2) 7
DK .... (SKIP TO JD2) 8

C. Does your airline require the terrain display to be selected during descent and landing?

NO OR NEVER 0
YES OR SOMETIMES 1
RF 7
DK 8

D. For times that terrain display is not required, do you usually use it during takeoff?

NO, NOT USUALLY 0
YES, USUALLY 1
RF 7
DK 8

E. For times that terrain display is not required, do you usually use it during descent and landing?

NO, NOT USUALLY 0
YES, USUALLY 1
RF 7
DK 8

F. Has the terrain display experienced a map shift on any aircraft on which you were a crew member?

NO OR NEVER 0
YES OR SOMETIMES 1
RF 7
DK 8

JD2. During the last 60 days, how many times did an aircraft on which you were a crewmember experience a ground proximity warning?

TIME .... |_|_|_|

IF ZERO, SKIP TO JD3

A. Was (this warning/ the most recent of these warnings) valid?

NO .... (SKIP TO JD3) 0
YES 1
RF .... (SKIP TO JD3) 7
DK .... (SKIP TO JD3) 8

B. During this (most recent) warning, did you see the approaching terrain on the terrain display before you heard the aural warning?

NO .... (SKIP TO JD3) 0
YES 1
RF .... (SKIP TO JD3) 7
DK .... (SKIP TO JD3) 8

JD3. During the last 60 days, how many times did which you were a crewmember receive a Minimum Safe Altitude Warning Alert, also known as an MSAW (em-saw) or an altitude awareness call from an A.T.C controller?

TIME .... |_|_|_|

IF ZERO, SKIP TO JD4

A. (During the most recent of these events,) What did your aircraft do in response to the warning?

____

____

____

B. (During this most recent A.T.C. warning event,) Did the aircraft have an enhanced G.P.W.S. or T.A.W.S. (taws) installed?

NO .... (SKIP TO JD4) 0
YES 1
RF .... (SKIP TO JD4) 7
DK .... (SKIP TO JD4) 8

GPWS = GROUND PROXIMITY WARNING SYSTEM
TAWS = TERRAIN AVOIDANCE WARNING SYSTEM

1. Did your aircraft also receive a ground proximity warning from this system?

NO 0
YES 1
RF 7
DK 8

JD4. How many times in the last 60 days, did an aircraft on which you were a crewmember fly a non-precision approach?

TIME .... |_|_|_|

IF ZERO, SKIP TO JD8

A. (Was this non-precision approach flown in I.M.C? / How many of these non-precision approaches were flown in I.M.C?)

TIME .... |_|_|_|

IMC = INSTRUMENT METEOROLOGICAL CONDITIONS

JD5. How many times in the last 60 days did an aircraft on which you were a crewmember fly an un-stabilized non-precision approach where the aircraft was not in landing configuration, on airspeed, or on glide-slope by 1,000 feet I.M.C or 500 feet V.M.C?

MC = METEOROLOGICAL CONDITIONS
VMC = VISUAL METEOROLOGICAL CONDITIONS

TIME .... |_|_|_|

IF ZERO, SKIP TO JD6

A. (During the most recent un-stabilized non precision approach,) What factors contributed to the inability to conduct a stabilized approach?

____

____

____

____

JD6. During the last 60 days, did an aircraft on which you were a crewmember have the choice between flying a constant angle approach or step-down non-precision approach?

NO .... (SKIP TO JD7) 0
YES 1
RF .... (SKIP TO JD7) 7
DK .... (SKIP TO JD7) 8

A. Which did you choose most often, the constant angle or step-down non-precision approach?

CONSTANT ANGLE 1
STEP-DOWN 2
CHOSE BOTH THE SAME 3
RF 7
DK 8

JD7. During the last 60 days, how many times did an aircraft on which you were a crewmember fly a non-precision approach to a runway when glide-slope information was available to you?

TIME .... |_|_|_|

IF ZERO, SKIP TO JD8

A. During (this/the most recent) non-precision approach, did you use the glide-slope information?

NO 0
YES 1
RF 7
DK 8

JD8. (Is the aircraft you fly/Are any of the aircraft you fly) LNAV/VNAV (L-nav/V-nav) capable?

LNAV = LATERAL NAVIGATION
VNAV = VERTICAL NAVIGATION

NO .... (SKIP TO JD9) 0
YES 1
RF .... (SKIP TO JD9) 7
DK .... (SKIP TO JD9) 8

A. Does your airline ever require pilots to use LNAV/VNAV (L-nav/V-nav) to fly constant angle approaches?

NO .... (SKIP TO JD9) 0
YES 1
RF .... (SKIP TO JD9) 7
DK .... (SKIP TO JD9) 8

1. In the last 60 days, how many times did an aircraft on which you were a crewmember use LNAV / VNAV (L-nav/V-nav) to fly constant angle approaches?

   # TIME .... |_|_|_|

B. During the last 60 days, how many times did an aircraft on which you were a crewmember not fly an LNAV/VNAV (L-nav/V-nav) approach when that option was available?

   # TIME .... |_|_|_|

   IF ZERO, SKIP TO JD9

1. Please explain why the LNAV/VNAV (L-nav/V-nav) approach wasn't flown (during the most recent time that it was available).

____

____

____

JD9. During the last 60 days, was an aircraft on which you were a crewmember equipped to meet Required Navigation Performance standards, sometimes called R.N.P?

NO	(SKIP TO JD10)	0
YES		1
RF	(SKIP TO JD10)	7
DK	(SKIP TO JD10)	8

A. Does your airline choose to use R.N.P?

NO	(SKIP TO JD10)	0
YES		1
RF	(SKIP TO JD10)	7
DK	(SKIP TO JD10)	8

B. How many times in the last 60 days did an aircraft on which you were a crewmember fly an R.N.P approach?

   # TIME .... |_|_|_|

C. During the last 60 days, how many times did any aircraft on which you were a crewmember not fly an R.N.P approach when that option was available?

   # TIME .... |_|_|_|

   IF ZERO, SKIP TO JD10

1. Please explain why the R.N.P. approach was not flown (most recent time that it was available).

____

____

____

JD10. **IF JD4 = 0, SKIP TO JD11**. During the last 60 days, how many times did an aircraft on which you were a crewmember fly a non-precision approach into an airport without D.M.E.?

TIME .... |_|_|_|

DME = DISTANCE MEASURING EQUIPMENT

IF ZERO, SKIP TO JD11

A. During (this event/the most recent of these events), would D.M.E have improved your ability to land safely?

NO 0
YES 1
RF 7
DK 8

JD11. During the last 60 days, how many times did an aircraft on which you were a crewmember fly an instrument approach into an airport where glide-slope or other ground based vertical angle guidance information was unavailable?

TIME .... |_|_|_|

IF ZERO, SKIP TO JD12

A. During (this approach/the most recent of these approaches), was D.M.E used to calculate the rate of descent for landing?

NO 0
YES 1
RF 7
DK 8

JD12. During the last 60 days, how many times did an aircraft on which you were a crewmember land on a runway without VASI (vasi) or PAPI (papi)?

TIME .... |_|_|_|

VASI = VERTICAL APPROACH SLOPE INDICATOR
PAPI = PRECISION APPROACH PATH INDICATOR

A. During the most recent of these events) would VASI (vasi) or PAPI (papi) have improved the aircraft's ability to land safely?

NO 0
YES 1
RF 7
DK 8

I would now like to ask you some questions about your airline's written standard operating procedures or SOPs.

JD13. Do your airline's written SOPs include Controlled Flight into Terrain prevention, sometimes called C-FIT (C-fit)?

NO 0
YES 1
RF 7
DK 8

JD14. Do your airline's written SOPs talk about how to avoid circumstances that could lead to an in-flight loss of control?

NO 0
YES 1
RF 7
DK 8

JD15. Do your airline's written SOPs talk about how to perform recovery from unusual **attitudes** and departure from controlled flight?

NO 0
YES 1
RF 7
DK 8

JD16. Do your airline's written SOPs talk about how to avoid approach and landing accidents?

NO 0
YES 1
RF 7
DK 8

JD17. Do your airline's written SOPs talk about how to fly non-precision approaches?

NO 0
YES 1
RF 7
DK 8

JD18. Do your airline's written SOPs require the use of constant angle non-precision approaches when that option is available?

NO 0
YES 1
RF 7
DK 8

JD19. Do your airline's written SOPs talk about how to respond to E.G.P.W.S warnings?

EGPWS = ENHANCED GROUND PROXIMITY WARNING SYSTEM

NO 0
YES 1
RF 7
DK 8

Now I would like to ask some questions about your recurrent training. By recurrent training I mean training conducted periodically that is designed to maintain your skills and knowledge. CLARIFICATION: This does not include transition or initial training. Recurrent training can include ground school, simulator training sessions, and any training conducted in the aircraft. I am going to read a list of issues. For each issue, please indicate if that topic or issue was covered during your last recurrent training.

JD20. In what month and year did you receive your most recent recurrent training?

MONTH .... |__|__|

YEAR .... |__|__|__|__|

JD21. Did your most recent recurrent training talk about basic airmanship?

NO 0
YES 1
RF 7
DK 8

A. Did your most recent recurrent training talk about normal approach procedures?

NO 0
YES 1
RF 7
DK 8

B. Did your most recent recurrent training talk about approach briefings?

NO 0
YES 1
RF 7
DK 8

C. Did your most recent recurrent training talk about criteria for initiating go-around and missed approaches?

NO 0
YES 1
RF 7
DK 8

D. Did your most recent recurrent training talk about go-around and missed approach execution?

NO 0
YES 1
RF 7
DK 8

E. Did your most recent recurrent training talk about emergency or abnormal conditions procedures?

NO 0
YES 1
RF 7
DK 8

Now I would like to ask you some questions concerning training you may have received addressing controlled flight into terrain, or C-FIT (C-fit), and other issues

JD22. Have you received C-FIT (C-fit) prevention training from your airline?

NO ....(SKIP TO JD23) 0
YES 1
RF ....(SKIP TO JD23) 7
DK ....(SKIP TO JD23) 8

A. In what month and year did you receive your most recent C-FIT (C-fit) prevention training?

MONTH .... |__|__|
YEAR .... |__|__|__|__|

B. Did your most recent C-FIT (C-fit) prevention training talk about minimum obstruction clearance altitudes or MOCA (mo ca)?

NO 0
YES 1
RF 7
DK 8

C. Did your most recent C-FIT (C-fit) prevention training talk about minimum enroute altitudes or M.E.A?

NO 0
YES 1
RF 7
DK 8

D. Did your most recent C-FIT (C-fit) prevention training talk about grid

MORA = MINIMUM OPERATING RADAR ALTITUDE

NO 0
YES 1
RF 7
DK 8

E. Did your most recent C-FIT (C-fit prevention training talk about G.P.W.S or E.G.P.W.S?

GPWS = GROUND PROXIMITY WARNING SYSTEM
EGPWS = ENHANCED GROUND PROXIMITY WARNING SYSTEM

NO 0
YES 1
RF 7
DK 8

F. Did your most recent C-FIT (C-fit) prevention training talk about escape maneuvers in response to G.P.W.S or G.P.W.S warnings?

GPWS = GROUND PROXIMITY WARNING SYSTEM
EGPWS = ENHANCED GROUND PROXIMITY WARNING SYSTEM

NO 0
YES 1
RF 7
DK 8

G. Did your most recent C-FIT (C-fit) prevention training talk about drift down procedures after engine failure?

NO 0
YES 1
RF 7
DK 8

H. Did your most recent C-FIT (C-fit) prevention training talk about maintaining situational awareness?

NO 0
YES 1
RF 7
DK 8

	I. Did your most recent C-FIT (C-fit) prevention training talk about cockpit resource management, or C.R.M as it relates to C-FIT (C-fit) recovery?	NO .... 0 YES .... 1 RF .... 7 DK 8
	NOTE: CRM CAN ALSO = <u>CREW</u> RESOURCE MANAGEMENT	
	J. How would you rate the quality of the most recent C-FIT (C-fit) prevention training you received from your airline? Would you say it was (READ CATEGORIES)?	EXCELLENT .... 1 GOOD .... 2 FAIR .... 3 POOR .... 4 VERY POOR 5
JD23.	Did you receive training specifically in upset recovery from your airline?	NO .... (SKIP TO JD24) .... 0 YES .... 1 RF .... (SKIP TO JD24) .... 7 DK .... (SKIP TO JD24) 8
	A. In what month and year did you receive your most recent training in upset recovery?	MONTH .... ⎣_⎦⎣_⎦ YEAR .... ⎣_⎦⎣_⎦⎣_⎦⎣_⎦
	B. Was this training received in a simulator, in a ground school, or both?	SIMULATOR .... 1 GROUND SCHOOL .... 2 BOTH .... 3 RF .... 7 DK 8
	C. How would you rate the quality of the upset recovery training you received? Would you say it was (READ CATEGORIES)?	EXCELLENT .... 1 GOOD .... 2 FAIR .... 3 POOR .... 4 VERY POOR 5
JD24.	Does your airline provide training in Cockpit or Crew Resource Management, sometimes called C.R.M?	NO .... (SKIP TO JD25) .... 0 YES .... 1 RF .... (SKIP TO JD25) .... 7 DK .... (SKIP TO JD25) 8
	A. Have you received this C.R.M training?	NO .... (SKIP TO JD25) .... 0 YES .... 1 RF .... (SKIP TO JD25) .... 7 DK .... (SKIP TO JD25) 8
	B. Did this C.R.M. training change how you manage the flight deck?	NO .... 0 YES .... 1 RF .... 7 DK 8
	C. Do you have suggestions for how the C.R.M training might be improved?	NO .... 0 YES .... 1 RF .... 7 DK 8
	D. What suggestions do you have?	

____

____

____

JD25. Does your airline have a no-fault missed approach or go-around policy?

NO ....(SKIP TO JD26) 0
YES 1
RF....................(SKIP TO JD26) 7
DK ....(SKIP TO JD26) 8

CLARIFICATION: No fault means that the airline does not apply disciplinary action or criticize pilots who exercise their authority to exercise a missed approach or go around.

A. Would you favor the institution of such policy, oppose it, or neither favor nor oppose it?

FAVOR 1
OPPOSE 2
NEITHER FAVOR NOR OPPOSE 3
RF.. 7
DK 8

JD26. During the last 60 days did you perform a missed approach or go around?

NO ....(SKIP TO JD27) 0
YES 1
RF....................(SKIP TO JD27) 7
DK ....(SKIP TO JD27) 8

A. Did you receive any feedback from your airline regarding this missed approach

NO ....(SKIP TO JD27) 0
YES 1
RF....................(SKIP TO JD27) 7
DK ....(SKIP TO JD27) 8

B. Was that feedback positive, negative, or both positive and negative?

POSITIVE 1
NEGATIVE 2
BOTH POSITIVE AND NEGATIVE 3
RF.. 7
DK 8

JD27. Does your airline participate in the safety reporting program called A-SAP (A-sap) also known as the Aviation Safety Action Program?

NO ....(SKIP TO JD28) 0
YES 1
RF....................(SKIP TO JD28) 7
DK ....(SKIP TO JD28) 8

A. Have you been briefed on this A-SAP (A-sap) program?

NO 0
YES 1
RF.. 7
DK 8

B. Were you told about the general purpose of the A-SAP (A-sap) program?

NO 0
YES 1
RF.. 7
DK 8

C. Were you told how to submit an A-SA A-sap) report?

NO 0
YES 1
RF.. 7
DK 8

D. If the situation arises in the future, would you submit an A-SAP (A-sap) report?

NO 0
YES ....(SKIP TO JD27E) 1
RF....................(SKIP TO JD27E) 7
DK .... (SKIP TO JD27E)8

1. Why not?

____

____

____

____

E. Do you believe that the confidentiality of A-SAP (A-sap) data is adequately protected?

NO		0
YES	(SKIP TO JD27E)	1
RF	(SKIP TO JD27E)	7
DK	(SKIP TO JD27E)	8

CLARIFICATION: Confidentiality refers to both the reporter and to the use of the data.

1. Why not?

____

____

____

____

F. Are you aware of any positive changes program other than A-SAP (A-sap) for receiving safety reports from pilots?

NO		0
YES		1
RF		7
DK		8

IF ZERO, SKIP TO JD29

JD28. Does your airline have a procedure or program other than A-SAP (A-sap) for receiving safety reports from pilots?

NO	(SKIP TO JD29)	0
YES		1
RF	(SKIP TO JD29)	7
DK	(SKIP TO JD29)	8

A. Are you aware of any positive changes that have resulted from this pilot reporting program?

NO		0
YES		1
RF		7
DK		8

B. Would you favor the establishment of an A-SAP (A-sap) program, oppose it, or neither favor nor oppose it?

FAVOR		1
OPPOSE		2
NEITHER FAVOR NOR OPPOSE		3
RF		7
DK		8

JD29. Does your airline have a Flight Operations Quality Assurance Program, sometimes called FOQA (FO Qua)?

NO	(ASK JD29A)	0
YES	(SKIP TO JD29B)	1
RF	(SKIP TO JD30)	7
DK	(SKIP TO JD30)	8

CLARIFICATION: This is a program at some airlines that analyzes operational data routinely collected from the flight data recorders with concurrence and oversight by the pilot's union or association at that airline.

A. Would you favor the establishment of a FOQA (FO Qua) program at your airline, oppose it, or neither favor nor oppose?

FAVOR		1
OPPOSE		2
NEITHER FAVOR NOR OPPOSE		3
RF		7
DK		8

IF ZERO, SKIP TO JD30

B. Have you been briefed on the program?	NO ... 0 YES ... 1 RF ... 7 DK ... 8
C. Do you believe that the confidentiality of FOQA (FO Qua) data is adequately protected? CLARIFICATION: Confidentiality refers to both the identity of the pilot flying the aircraft and to the use of the data.	NO ... 0 YES ... 1 RF ... 7 DK ... 8
D. Are you aware of any safety improvements that have resulted from the FOQA (FO Qua) program?	NO ... 0 YES ... 1 RF ... 7 DK ... 8

We are interested in hearing about the safety culture at your airline, as expressed by your senior management. By senior management, we mean the C.E.O., Director of Safety, V.P. for Safety, Director of Flight Operations, and other senior management.

CEO = CHIEF EXECUTIVE OFFICER
VP = VICE PRESIDENT

JD30.	Does your airline have a C.E.O. mission statement on safety? CEO = CHIEF EXECUTIVE OFFICER	NO ... 0 YES ... 1 RF ... 7 DK ... 8
JD31.	Does your airline have a Director of Safety?	NO ... 0 YES ... 1 RF ... 7 DK ... 8
JD32.	Does your airline have a V.P. of Safety? VP = VICE PRESIDENT	NO ... 0 YES ... 1 RF ... 7 DK ... 8
JD33.	Have you observed a strong commitment to safety among senior management? (This question focuses on behavior.)	NO ... (SKIP TO JD34) ... 0 YES ... 1 RF ... (SKIP TO JD34) ... 7 DK ... (SKIP TO JD34) ... 8
	A. Is this senior management commitment to safety reflected throughout the organization?	NO ... 0 YES ... 1 RF ... 7 DK ... 8
JD34.	If you have a safety concern, do you have a mechanism for bringing that concern to the attention of senior management?	NO ... (SKIP TO SECTION D) ... 0 YES ... 1 RF ... (SKIP TO SECTION D) ... 7 DK ... (SKIP TO SECTION D) ... 8
	A. How effective is this mechanism in reaching senior management? Would you say (READ CATEGORIES)?	EXTREMELY EFFECTIVE ... 1 VERY EFFECTIVE ... 2 SOMEWHAT EFFECTIVE ... 3 NOT VERY EFFECTIVE ... 4 NOT AT ALL EFFECTIVE ... 5

Air Carrier Questionnaire

Section D: Questionnaire Feedback

SECTION D: QUESTIONNAIRE FEEDBACK

INTRODUCTION:
I only have a couple more questions and these are about your reactions to the survey we have just done.

D1. How confident are you that you accurately counted all of the safety-related events that I asked you about? Would you say you were (READ QUESTIONS)?

- Not confident at all.......... 1
- Slightly confident.......... 2
- Moderately confident.......... 3
- Very confident.......... 4
- Extremely confident.......... 5
- RF.......... 7
- DK.......... 8

D2. Were any of the questions I asked confusing, poorly worded, or ambiguous?

- YES.......... 1
- NO..........(SKIP TO D3).......... 0
- RF..........(SKIP TO D3).......... 7
- DK`..........(SKIP TO D3).......... 8

A. Could you please describe these question problems? RECORD VERBATIM. AT COMPLETION OF INTERVIEW, ENTER QUESTION NUMBER.

QUESTION NUMBER	RECORD VERBATIM

D3. Are there any safety problems happening within the national aviation system that I did not ask about but that you think may be worth asking about in further surveys?

- YES.......... 1
- NO..........(SKIP TO D4).......... 0
- RF..........(SKIP TO D4).......... 7
- DK`..........(SKIP TO D4).......... 8

A. What are these problems?

SPECIFY: ______________________________

D4. Do **you** use the internet at home?

- YES.......... 1
- NO.......... 0
- RF.......... 7
- DK.......... 8

D5. Do you have any other comments or suggestions about this survey? RECORD VERBATIM.

____

____

____

____

PANEL PASSWORD HINT	TAKES INTERVIEWER TO "NEEDPAS" (PANEL 1ST QTR OR LATER QTR BUT NEVER COMPLETED INTERVIEW) OR PAST PATH (PANEL 2ND QTR OR LATER WHO PREVIOUSLY GAVE PASSWORD).
NEEDPASS: We would like to be able to link the information you give us each time we call. Because we do not link your information with your name, we would like to record an individual password we can use to link your data. May we please have a password that you will repeat to us when we call you again?	AGREED .... 1 REFUSED .... (ENDINT) 7
PICKPASS: RECORD PASSWORD	TAKES INTERVIEWER TO ENDINT.
ASKFORHINT: Please give us a question that we can use as a hint in case you are unable to remember your password the next time we call. For instance, if you choose the word "RED" as your password, your hint question could be "What is my favorite color?"	RECORD HINT
PASTPATH: At the end of your last interview you gave us a password so we could link your information across quarters. Your hint question was (HINTQUESTION). What was your password? RECORD.	REMEMBERS PASSWORD .... (REPREATPASS) .... 1 REFUSED .... (ENDINT) .... 7 CAN'T REMEMBER .... (SUBSPASS) 8
REPEATPASS: RECORD PASSWORD.	IF SUCCESSFUL, TAKES INTERVIEWER TO ENDINT.
IF PASSWORD NOT IN PASSWORD LIST: The word you gave me does not match our list of passwords. Perhaps I spelled it wrong. How do you spell your password? RETURN TO REPEATPASS FIELD AND RECORD PASSWORD AGAIN. IF WORD STILL DOESN'T MATCH AFTER TWO ATTEMPTS, CLICK, SUPPRESS.	IF SUPPRESSED, TAKES INTERVIEWER TO SUBSPASS.

SUBSPASS: Since (you can't remember/we don't seem to have) your previous password, we'd like you to choose another password and hint so we can link your future interviews. May we please have another password and hint that you will repeat to us when we call again?	YES .... (PICKPASS) .... 1 NO .... (ENDINT) 0

ENDINT Again, thank you very much for your time and your help with this survey. Your input will help the aviation industry a great deal to measure the level of safety in the aviation system and will be held in confidence.

IF PANEL MEMBER: We'll be calling again in three months for your (2nd/3rd/last) interview.

QUESTIONNAIRE LENGTH: QUESTIONNAIRE LENGTH (MINUTES) .... |__|__|__|

H

General Aviation Questionnaire

The following excerpt from the *NAOMS Reference Report* (Battelle, 2007) introduces the general aviation questionnaire, which is reprinted here.

> NAOMS began interviewing general aviation (GA) pilots on August 7th, 2002. The approach used for GA pilot interviews was quite similar to that used for air carrier (AC) pilots. The questionnaire consisted of four sections that corresponded with the general topics covered in the air carrier questionnaire: Section A addressed pilot qualifications and experience; Section B addressed safety events; Section C addressed a specific focus topic (weather-related issues); and Section D offered pilots an opportunity to provide feedback on the interview process and the questionnaire. This appendix contains a copy of the GA questionnaire.

Appendix 12: General Aviation Questionnaire

INTRODUCTION:	Hello, I'm (NAME) calling for NASA. We sent you a letter a few days ago to tell you about the National Aviation Operations Monitoring Service project that NASA is conducting.

SQ1. Did you receive the letter?

NO ....(READ SCRIPT).................................... 0
YES ....(SKIP TO SQ2)..................................... 1

SCRIPT: I'm sorry. Let me read to you what it says. The purpose of this project is to provide reliable safety data for improving aviation safety.

We randomly selected pilots from the pilot registry to participate in this interview. The interview will only take about 30 minutes. The anonymous information you provide will be combined with information submitted by about 10,000 other general aviation pilots. Your participation is entirely voluntary. The information you provide will be held in complete confidence. NASA **will not** retain any record of your identity, or link your name to your answers in any way.

SQ2. Before beginning, first let me check your recent flight experience against our survey requirements. Have you flown as a pilot or co-pilot logging hours on an airplane or in a helicopter during the last 60 days? Please do not include non-powered aircraft, military or ultra-light flying. (Note to interviewers: "co-pilot logging hours means that you flew as a co-pilot and logged hours in your official FAA logbook.")

NO ....(SKIP TO TERMINATION SCRIPT)...................... 0
YES 1

TERMINATION SCRIPT: I'm sorry, but we are only interviewing pilots who have flown in the last 60 days. NASA would like to thank you for your time. Goodbye. **TERMINATE INTERVIEW**

SQ3. The interview may be monitored by my supervisor for quality control purposes. Is this a good time to proceed?

NO .... (FILL OUT CALL BACK INFORMATION) 0
YES .... (PROCEED TO A1) 1

TIME BEGUN .... (MILITARY)................ |__|__|:|__|__|
(FILLS)

INTERVIEWER: DATE OF INTERVIEW IS BEING RECORDED AS (START DATE).
IS THIS THE CORRECT DATE:

NO.................(RECORD DATE OF INTERVIEW)................ 0
YES 1

START DATE ....|__|__|/|__|__|/|__|__|
MONTH DAY YEAR

.... **START DATE = 60 DAYS BEFORE END DATE**

END DATE...|__|__|/|__|__|/|__|__|
(FILLS) MONTH DAY YEAR

END DATE = DAY BEFORE DAY OF INTERVIEW

General Aviation Questionnaire

Section A:
Background Questions

SECTION A: BACKGROUND QUESTIONS

INTRODUCTION: The first few questions are about your general flying experience.

	Question	Response	Code
A1	Do you hold an A.T.P certificate or instrument rating? ATP=AIRLINE TRANSPORT PILOT	NO	0
		YES	1
		RF	7
		DK	8
	A. Are you I.F.R current? NOTE: IFR = Instrument Flight Rules	NO	0
		YES	1
		RF	7
		DK	8
A2	During your **life,** approximately how many **hours in total** have you flown as a pilot? Include all types of flying including FAR part 121 air carrier, military service, and ultralight flying.	TOTAL HOURS DURING LIFE	__,___
		RF	99 997
		DK	99 998

INTRODUCTION: The rest of the questions will refer to 60 days prior to today. Whenever I say the "last 60 days," I am referring to the period from (START DATE) through (END DATE). Also, for all these questions, I will be asking you about events when you flew as a pilot in command or copilot logging hours in your official FAA logbook under FAR Part 121, Part 135 or Part 91. First I would like to ask a few questions about the type of flying you have done in the last 60 days.

	Question	Response	Code
A3.	During the last 60 days, how many **hours** did you fly as a pilot or copilot under FAR Part 121, Part 135, or Part 91?	TOTAL HOURS FLOWN LAST 60 DAYS	___
		NO HOURS(TERMINATE INTERVIEW, CODE "NOT ELIGIBLE"	
		IIF HOURS IN A3 ARE ABOVE 300 ASK A. OTHER RESPONSES SKIP TO A4	
	A. I'd just like to verify. You said you flew (# HOURS A3) **hours** during the last 60 days. Is this correct?	NO (ASK B)	0
		YES (SKIP TO A4)	1
		RF (SKIP TO A4)	7
		DK (SKIP TO A4)	8
	B. During the last 60 days, how many **hours** did you fly? NOTE TO INTERVIEWER: AS A PILOT OR COPILOT UNDER FAR PART 121, PART 135 OR PART 91.	# HOURS	___
		RF	997
		DK	998
A4	How many of these (HOURS IN A3) hours did you fly as an airplane pilot or copilot **under FAR Part 121**?	# OF HOURS FAR PART 121	___
		RF	997
		DK	998
		HOURS CANNOT EXCEED HOURS IN A3.	
A5	How many of these (HOURS IN A3) hours did you fly as an pilot or copilot **under FAR Part 135?**	# HOURS UNDER FAR 135	___
		RF	997
		DK	998
		HOURS CANNOT EXCEED HOURS IN A3 MINUS A4. IF >0, ASK A. OTHERS SKIP TO A6.	
	A. Of the (HOUR IN A5) hours flown under Part 135, how many were flown on fixed wing airplanes?	# HOURS FAR 135 AIRPLANE	___
		RF	997
		DK	998
		HOURS CANNOT EXCEED HOURS IN A5. IF = A5, SKIP TO A6. IF <A5, ASK A5B.	

B. Of the (HOUR IN A5) hours flown under Part 135, how many were flown on helicopters?

HOURS FAR 135 HELICOPTER ....
RF 997
DK 998
HOURS CANNOT EXCEED HOURS IN A5 MINUS A5A.

A6 How many of these (HOURS IN A3) hours did you fly as a pilot or copilot **under FAR Part 91?**

HOURS UNDER FAR 91 ....
RF 997
DK 998
HOURS CANNOT EXCEED HOURS IN A3 MINUS SUM (A4 PLUS A5).
IF >0, ASK A. OTHERS SKIP TO A6.

A. Of the (HOUR IN A65) hours flown under Part 91, how many were flown on fixed wing airplanes?

HOURS FAR 91 AIRPLANE ....
RF 997
DK 998
HOURS CANNOT EXCEED HOURS IN A6.
IF = A6, SKIP TO A7.
IF <A6, ASK A6B.

B. Of the (HOUR IN A5) hours flown under Part 91, how many were flown on helicopters?

HOURS FAR 91 HELICOPTER ....
RF 997
DK 998
HOURS CANNOT EXCEED HOURS IN A6 MINUS A6A.

INTRODUCTION: Now I'd like to ask a few questions about the number of takeoffs or flights you made during the last 60 days. We use the terms "flight" throughout this interview to mean the period of time between each takeoff and landing, even if that time is short such as for instructors teaching students to land. READ A7-A11 WHEN APPLICABLE.

A7 IF A4 > 0, READ: During the last 60 days you mentioned you flew (# HOUR A4) hours as an **airplane pilot or copilot under FAR Part 121.** How many flights or legs did you experience as a pilot or copilot during this period?

OF LEGS/TAKEOFFS PART 121 ....
RF 997
DK 998

A8 IF A5A > 0, READ: During the last 60 days you mentioned you flew (# HOUR A5A) hours as an **airplane** pilot or copilot **under FAR Part 135.** How many **takeoffs** did you experience as an airplane pilot or copilot during this period?

PART 135 AIRPLANE TAKEOFFS ....
RF 997
DK 998

IF A8 BLANK OR 0, SKIP TO A9.

A. Of these (# TAKEOFFS A8) flights, how many occurred during night time conditions?

PART 135 AIRPLANE FLIGHTS NIGHT ....
RF 997
DK 998

MUST BE EQUAL TO OR LESS THAN A8

B. Of these (# TAKEOFFS A8) flights, how many were at least 50 nautical miles long?

PART 135 AIRPLANE FLIGHTS LONG ....
RF 997
DK 998

MUST BE EQUAL TO OR LESS THAN A8

A9 IF A5B > 0, READ: During the last 60 days you mentioned you flew (# HOUR A5A) hours as **a helicopter** pilot or copilot **under FAR Part 135.** How many **takeoffs** did you experience as a helicopter pilot or copilot during this period?

PART 135 HELICOPTER TAKEOFFS ....
RF 997
DK 998

IF A9 BLANK OR 0, SKIP TO A10.

A. Of these (# TAKEOFFS A9) flights, how many occurred during night time conditions?

PART 135 HELICOPTER FLIGHTS NIGHT .... |_|_|_|
RF 997
DK 998

MUST BE EQUAL TO OR LESS THAN A9

B. Of these (# TAKEOFFS A9) flights, how many were at least 50 nautical miles long?

PART 135 HELICOPTER FLIGHTS LONG .... |_|_|_|
RF 997
DK 998

MUST BE EQUAL TO OR LESS THAN A9

A10 D. IF A6A > 0, READ: During the last 60 days you mentioned you flew (# HOUR A5A) hours as **an airplane** pilot or copilot **under FAR Part 91.** How many **takeoffs** did you experience as an airplane pilot or copilot during this period?

OF TAKEOFFS PART 91 AIRPLANE .... |_|_|_|
RF 997
DK 998

IF A10 BLANK OR 0, SKIP TO A11.

A. Of these (# TAKEOFFS A10) flights, how many occurred during night time conditions?

PART 91 AIRPLANE FLIGHTS NIGHT .... |_|_|_|
RF 997
DK 998

MUST BE EQUAL TO OR LESS THAN A10

B. Of these (# TAKEOFFS A9) flights, how many were at least 50 nautical miles long?

PART 135 AIRPLANE FLIGHTS LONG .... |_|_|_|
RF 997
DK 998

MUST BE EQUAL TO OR LESS THAN A10

A11 IF A6B > 0, READ: During the last 60 days you mentioned you flew (# HOUR A5A) hours as a **helicopter** pilot or copilot **under FAR Part 91.** How many **takeoffs** did you experience as a helicopter pilot or copilot during this period?

OF TAKEOFFS PART 91 HELICOPTER .... |_|_|_|
RF 997
DK 998

IF A11 BLANK OR 0, SKIP TO A12.

A. Of these (# TAKEOFFS A10) flights, how many occurred during night time conditions?

PART 91 HELICOPTER FLIGHTS NIGHT .... |_|_|_|
RF 997
DK 998

MUST BE EQUAL TO OR LESS THAN A11

B. Of these (# TAKEOFFS A9) flights, how many were at least 50 nautical miles long?

PART 135 HELICOPTER FLIGHTS LONG .... |_|_|_|
RF 997
DK 998

MUST BE EQUAL TO OR LESS THAN A11

A12 ASK ONLY IF A1 = YES. During the last 60 days, for how many flights did you file an I.F.R flight plan?

FILED IFR FLIGHT PLAN .... |_|_|_|
RF 997
DK 998

FLIGHTS CANNOT EXCEED SUM OF (A8+A9+A10+A11)

INSTRUCTION:

REVIEW A1, A2 AND A3, THEN FOLLOW THE ***FIRST*** INSTRUCTION THAT APPLIES.

- IF A1 HIGHEST, FOLLOW AIRPLANE ROUTE.
- IF A2 HIGHEST, FOLLOW AIR CARRIER ROUTE.
- IF A3 HIGHEST, FOLLOW HELICOPTER ROUTE.
- IF ANY TWO OR THREE ARE EQUAL, FOLLOW ROUTES IN FOLLOWING PRIORITY:
  - HELICOPTER
  - AIRPLANE
  - AIR CARRIER
- ***OR ANY OTHER ORDER WE WOULD PREFER!***

INTRODUCTION:

You mentioned that during the last 60 days you flew (# HOURS DECIDED FROM ABOVE) hours flying as (an airplane/a helicopter) pilot or copilot under FAR (Part 135/Part 91/Part 135 and Part 91). For the rest of the interview, I will be asking you about your experiences flying (airplanes/helicopters) during this period of time.

A13 AIRPLANES AND HELICOPTERS During the last 60 days, did you fly an airplane engaged in FAR Part 135 or Part 91 air carrier operations in any of the following capacities (READ CATEGORIES)?

	YES	NO	RF	DK
a. as a captain or pilot in command..........................	1	0	7	8
b. as a copilot or first officer	1	0	7	8
c. as an instructor pilot	1	0	7	8
d. in any other capacity	1 (ASK 1)	0	7	8
1. What was that capacity? SPECIFY				

SPECIFY:______________________________

A14. I am now going to read a list of different types of general aviation flying. Please tell me if you engaged in any of these types of flying during the last 60 days. Did you undertake any (airplane/helicopter) flights (READ CATEGORIES)?

	NO	YES ***(ASK COL I.)***	RF	DK	***COL I.*** Approximately how many hours would you say was devoted to (BOLD WORDS IN A9a-g)?
a. for **flight instruction as the instructor** .... (NOTE: INCLUDES CHECKOUT FLIGHTS)	0	1	7	8	|_|_|_|
b. for **flight instruction as the student**..... (NOTE: INCLUDES CHECKOUT FLIGHTS)	0	1	7	8	|_|_|_|
c. for **corporate transportation** as a pilot employee of a corporate flight department... (NOTE: DOES NOT INCLUDE CHARTER FLIGHTS)	0	1	7	8	|_|_|_|
d. for you to **support your own business**..	0	1	7	8	|_|_|_|

e.	in **aircraft owned or operated by government entities**, sometimes called public use flights	0	1	7	8	|_|_|_|
f.	**flights with revenue passengers**? (NOTE: THIS MEANS PAYING PASSENGERS WHO ARE ON THE AIRCRAFT)	0	1	7	8	|_|_|_|
g.	**flights that carried only cargo or freight and did not carry revenue passengers**?	0	1	7	8	|_|_|_|
f.	for **transporting patients or critical medical products** such as organs for transplant or blood	0	1	7	8	|_|_|_|
g.	for **recreation or personal transportation** not associated with business	0	1	7	8	|_|_|_|
h.	for any other purpose (SPECIFY)	0	1	7	8	|_|_|_|

SPECIFY VERBATIM: ______________________________

A15. Please tell me all of the (airplane/helicopter) makes and models you flew as a pilot or copilot during the last 60 days. RECORD VERBATIM. **LIST ALL MODELS THEN ASK COL I. AND II. FOR EACH.**

MAKE/MODEL (RECORD VERBATIM)	*COL I.* During the last 60 days, how many hours you fly the (MAKE/MODEL)?	*COL II.* How many engines does this aircraft have?	COL III Is this an experimental airplane?
1st ____________	|_|_|_|	|_|_|_|	YES NO RF DK
2nd ____________	|_|_|_|	|_|_|_|	YES NO RF DK
3rd ____________	|_|_|_|	|_|_|_|	YES NO RF DK
4th ____________	|_|_|_|	|_|_|_|	YES NO RF DK
5th ____________	|_|_|_|	|_|_|_|	YES NO RF DK
6th ____________	|_|_|_|	|_|_|_|	YES NO RF DK

A16 (AIRPLANE ONLY) During the last 60 days, how many hours did you fly an experimental airplane?

HOURS FLY EXPERIMENTAL..............
RF..........(SKIP TO a14).............997
DK..........(SKIP TO A14).............998

A. What was the make and model of the experimental airplane?

RECORD MAKE/MODEL

General Aviation Questionnaire

Section B:
Safety Related Events

SECTION B: SAFETY RELATED EVENTS

INTRODUCTION:

My next set of questions are about safety related events. Just as a reminder, I'd like you to report only events that **you experienced flying under FAR Part 135 or Part 91 on (an airplane/a helicopter) on which you were a** pilot or copilot. The first questions are about **equipment-related events.**

ER1. How many times during the last 60 days did (an airplane/a helicopter) on which you were a pilot or copilot divert to an alternate airport or return to land because of (an airplane/ a helicopter) equipment problem?

EQUIPMENT PROBLEMS .... |__|__|__|

IF 0, SHIP TO ER2.

A. ASK ONLY IF MORE THAN ONE MAKE/ MODEL IN A10. Which (airplane/helicopter) experienced this equipment problem? Was it (READ A10 MAKE/MODEL LIST).

RECORD MAKE/MODEL

B. What systems caused the diversion or return to land?

SPECIFY:

ER2-A AIRPLANES ONLY

I am going to read a list of possible airplane malfunctions or failures. For each one, please tell me how many times during the last 60 days an **in-flight airplane** on which you were a pilot or copilot experienced any of these malfunctions or failures. If a piece of equipment does not apply, please answer "not applicable" rather than "zero". How many times did you experience (READ QUESTIONS):

A. Uncommanded movements of the speedbrakes?

SPEEDBRAKERS .... |__|__|__|

IF 0, SKIP TO B.

1. ASK ONLY IF MORE THAN ONE MAKE/ MODEL IN A10.

Which aircraft experienced this malfunction? Was it (READ A10 MAKE/MODEL LIST)?

SPECIFY MAKE/MODEL:

B. Uncommanded movements of the trim tabs?

TRIM TABS.. |__|__|__|

IF 0, SKIP TO C.

1. ASK ONLY IF MORE THAN ONE MAKE/MODEL IN A10.

Which aircraft experienced this malfunction? Was it (READ A10 MAKE/MODEL LIST)?

SPECIFY MAKE/MODEL:

C. Uncommanded movements of the flaps? — # FLAPS .... |_|_|_|
IF 0, SKIP TO D.

1. ASK ONLY IF MORE THAN ONE MAKE/MODEL IN A10.

Which aircraft experienced this malfunction? Was it (READ A10 MAKE/MODEL LIST — SPECIFY MAKE/MODEL:

D. Failure of the trim system to operate? — # TRIM .... |_|_|_|
IF 0, SKIP TO E.

1. ASK ONLY IF MORE THAN ONE MAKE/MODEL IN A10.

Which aircraft experienced this failure? Was it (READ A10 MAKE/MODEL LIST)? — SPECIFY MAKE/MODEL:

E. Failure of the landing gear to extend or retract? — # GEAR .... |_|_|_|
IF 0, SKIP TO F.

1. ASK ONLY IF MORE THAN ONE MAKE/MODEL IN A10.

Which aircraft experienced this failure? Was it (READ A10 MAKE/MODEL LIST)? — SPECIFY MAKE/MODEL:

F. Failure of the flaps to extend or retract? — # FLAPS IN OR OUT .... |_|_|_|
IF 0, SKIP TO G.

1. ASK ONLY IF MORE THAN ONE MAKE/MODEL IN A10.

Which aircraft experienced this failure? Was it (READ A10 MAKE/MODEL LIST)? — SPECIFY MAKE/MODEL:

G. Did you experience a malfunction or failure of any other aircraft device or system during the last 60 days?

YES		1
NO	(SKIP TO ER3)	0
RF	(SKIP TO ER3)	7
DK	(SKIP TO ER3)	8
NA	(SKIP TO ER3)	9

Which aircraft experienced this malfunction or failure? Was it (READ A10 MAKE/MODEL LIST)? — SPECIFY MAKE/MODEL:

Which device or system malfunctioned or failed?

SPECIFY: ______________________________

ER2-H HELICOPTER ONLY.
I am going to read a list of possible helicopter malfunctions or failures. For each one, please tell me how many times during the last 60 days an **in-flight helicopter** on which you were a pilot or copilot experienced any of these malfunctions or failures. If a piece of equipment does not apply, please answer "not applicable" rather than "zero". How many times did you experience (READ QUESTIONS):

A. Uncommanded movements of the trim? # UNCOMMANDED TRIM .... |__|__|__|
IF 0, SKIP TO B.

1. ASK ONLY IF MORE THAN ONE MAKE/ MODEL IN A10.

Which helicopter experienced this malfunction? Was it (READ A10 MAKE/MODEL LIST)? SPECIFY MAKE/MODEL:

B. Failure of the trim system to operate? # FAILURE TRIM .... |__|__|__|
IF 0, SKIP TO C.

1. ASK ONLY IF MORE THAN ONE MAKE/ MODEL IN A10.

Which helicopter experienced this failure? Was it (READ A10 MAKE/MODEL LIST)? SPECIFY MAKE/MODEL:

C. Failure of the landing gear to extend or retract? # FAILURE GEAR .... |__|__|__|
IF 0, SKIP TO D.
N/a

1. ASK ONLY IF MORE THAN ONE MAKE/ MODEL IN A10. SPECIFY MAKE/MODEL:

Which helicopter experienced this failure? Was it (READ A10 MAKE/MODEL LIST)?

D. Tail rotor failure? # FAILURE TAIL ROTOR .... |__|__|__|
IF 0, SKIP TO E.

1. ASK ONLY IF MORE THAN ONE MAKE/ MODEL IN A10.

SPECIFY MAKE/MODEL:
Which helicopter experienced this failure? Was it (READ A10 MAKE/MODEL LIST)?

E. Failure of hydraulic system?

FAILURE HYDRAULIC SYSTEM .... |_|_|_|
IF 0, SKIP TO F.

1. ASK ONLY IF MORE THAN ONE MAKE/ MODEL IN A10.

   Which helicopter experienced this failure? Was it (READ A10 MAKE/MODEL LIST)?

   SPECIFY MAKE/MODEL:

F. Valid transmission warning of potential failure?

TRANSMISSION WARNING .... |_|_|_|
IF 0, SKIP TO G.

1. ASK ONLY IF MORE THAN ONE MAKE/ MODEL IN A10.

   Which helicopter experienced this warning? Was it (READ A10 MAKE/MODEL LIST)?

   SPECIFY MAKE/MODEL:

G. Did you experience a malfunction or failure of any other aircraft device or system during the last 60 days?

YES 1
NO .... (SKIP TO ER3) 0
RF .... (SKIP TO ER3) 7
DK .... (SKIP TO ER3) 8

1. ASK ONLY IF MORE THAN ONE MAKE/ MODEL IN A10.

   Which helicopter experienced this malfunction or failure? Was it (READ A10 MAKE/MODEL LIST)?

   SPECIFY MAKE/MODEL:

   1) Which device or system malfunctioned or failed? ______________________

ER3. How many times during the last 60 days did an inflight (airplane/helicopter) on which you were a pilot or copilot experience smoke, fire, or fumes that originated in any of the following areas (READ QUESTIONS):

A. the engine, engine compartment or nacelle?

IN ENGINE OR NACELLE .... |__|__|__|

IF 0, SKIP TO ER3B.

1. ASK ONLY IF MORE THAN ONE MAKE/MODEL IN A10.

   Which (airplane/helicopter) experienced smoke, fire or fumes in the engine or nacelle? Was it (READ A10 MAKE/MODEL LIST)?

   SPECIFY MAKE/MODEL:

2. (Of the [# in ER3A] times there was smoke, fire, or fumes in the engine or nacelle, how many involved/Did the smoke, fire, or fumes in the engine or nacelle involve) electrical components or wiring?

   # NACELLE SMOKE/FIRE/FUMES .... |__|__|__|

   THE AMOUNT IN ER3A1 CANNOT BE GREATER THAN THE AMOUNT IN ER3A.

B. the cockpit?

IN COCKPIT .... |__|__|__|

IF 0, SKIP TO ER3C

1. ASK ONLY IF MORE THAN ONE MAKE/MODEL IN A10.

   Which (airplane/helicopter) experienced smoke, fire or fumes in the cockpit? Was it (READ A10 MAKE/MODEL LIST)?

   SPECIFY MAKE/MODEL:

2. (Of the [# in ER3B] times there was smoke, fire, or fumes in the cockpit, how many involved/Did the smoke, fire, or fumes in the cockpit deck involve) electrical components or wiring?

   COCKPIT SMOKE/FIRE/FUMES .... |__|__|__|

   THE AMOUNT IN ER3B1 CANNOT BE GREATER THAN THE AMOUNT IN ER3B.

C. the cargo or baggage area? — # IN CARGO AREA.................. |__|__|__|

IF 0, SKIP TO ER3D.

1. ASK ONLY IF MORE THAN ONE MAKE/MODEL IN A10.

   Which (airplane/helicopter) experienced smoke, fire or fumes in the cargo or baggage area? Was it (READ A10 MAKE/MODEL LIST)?

   SPECIFY MAKE/MODEL:

2. (Of the [# in ER3C] times there was smoke, fire, or fumes in the cargo area, how many involved/Did the smoke, fire, or fumes in the cargo area involve) electrical components or wiring?

   CARGO SMOKE/FIRE/FUMES.................. |__|__|__|

   THE AMOUNT IN ER3C1 CANNOT BE GREATER THAN THE AMOUNT IN ER3C.

D. REPEAT INTRODUCTION: the passenger compartment area? — # IN ELECTRICAL PASSENGER AREA.................. |__|__|__|

IF 0, SKIP TO ER3E.

1. ASK ONLY IF MORE THAN ONE MAKE/MODEL IN A10.

   Which (airplane/helicopter) experienced smoke, fire or fumes in the passenger compartment area? Was it (READ A10 MAKE/MODEL LIST)?

   SPECIFY MAKE/MODEL:

2. (Of the [# in ER3D] times there was smoke, fire, or fumes in the passenger compartment, how many involved/Did the smoke, fire, or fumes elsewhere in the passenger compartment involve) electrical components or wiring?

   SMOKE/FIRE/FUMES.................. |__|__|__|

   THE AMOUNT IN ER3D1 CANNOT BE GREATER THAN THE AMOUNT IN ER3D.

E. How many times (an airplane/a helicopter) experience smoke, fire or fumes that originated someplace other than in the engine or nacelle, cockpit, cargo area, or passenger area? — # ORIGINATE OTHER PLACES.................. |__|__|__|

IF 0, SKIP TO ER4.

1. ASK ONLY IF MORE THAN ONE MAKE/MODEL IN A10.

   Which (airplane/helicopter) experienced smoke, fire or fumes originating elsewhere? Was it (READ A10 MAKE/MODEL LIST)?

   SPECIFY MAKE/MODEL:

2. Where did the smoke, fire or fumes originate? SPECIFY.

   SPECIFY: ______________________________

ER4.	During the last 60 days, how many times did an inflight (airplane/helicopter) on which you were a pilot or copilot experience a precautionary engine shutdown?	# PRECAUTIONARY ENGINE SHUTDOWNS............ |__|__|__| **IF 0, SKIP TO ER5.**
	A. ASK ONLY IF MORE THAN ONE MAKE/MODEL IN A10. Which (airplane/helicopter) experienced a precautionary engine shutdown? Was it (READ A10 MAKE/MODEL LIST)?	SPECIFY MAKE/MODEL:
ER5.	Experience a total engine failure?	# TOTAL ENGINE FAILURE.. |__|__|__| **IF 0, SKIP TO ER6.**
	A. ASK ONLY IF MORE THAN ONE MAKE/MODEL IN A10. Which (airplane/helicopter) experienced a total engine failure? Was it (READ A10 MAKE/MODEL LIST)?	SPECIFY MAKE/MODEL:
ER6.	Experience total loss of electrical power?	# TOTAL ELECTRICAL FAILURE.............................. |__|__|__| **IF 0, SKIP TO ER7.**
	A. ASK ONLY IF MORE THAN ONE MAKE/MODEL IN A10. Which (airplane/helicopter) experienced a total loss of electrical power? Was it (READ A10 MAKE/MODEL LIST)?	SPECIFY MAKE/MODEL:
ER7.	During the last 60 days when you were pilot or copilot, how many times did you discover that your (airplane/helicopter) had incorrect or bogus parts installed?	# TOTAL PARTS.. |__|__|__| **IF 0, SKIP TO ER8.**
	A. ASK ONLY IF MORE THAN ONE MAKE/MODEL IN A10. Which (airplane/helicopter) had incorrect or bogus parts installed? Was it (READ A10 MAKE/MODEL LIST)?	SPECIFY MAKE/MODEL:

ER8.	Have cabin doors, baggage doors or cowlings open inadvertently during flight ?	# TOTAL DOORS OPEN .... |_|_|_| **IF 0, SKIP TO ER9.**
	A. ASK ONLY IF MORE THAN ONE MAKE/MODEL IN A10. Which (airplane/helicopter) had doors or cowlings open inadvertently during flight? Was it (READ A10 MAKE/MODEL LIST)?	SPECIFY MAKE/MODEL:
ER9.	Have a door or window come off of the aircraft while in flight?	# TOTAL DOORS OFF .... |_|_|_| **IF 0, SKIP TO ER10.**
	A. ASK ONLY IF MORE THAN ONE MAKE/MODEL IN A10. Which (airplane/helicopter) had doors or windows come off the (airplane/helicopter) while in flight? Was it (READ A10 MAKE/MODEL LIST)?	SPECIFY MAKE/MODEL:
ER10.	Experience a cargo shift or cargo coming loose?	# TOTAL CARGO LOOSE .... |_|_|_| **IF 0, SKIP TO ER11.**
	A. ASK ONLY IF MORE THAN ONE MAKE/MODEL IN A10. Which (airplane/helicopter) experienced a cargo shift or cargo coming loose? Was it (READ A10 MAKE/MODEL LIST)?	SPECIFY MAKE/MODEL:
ER11.	During the last 60 days, how many times did an (airplane/helicopter) on which you were a pilot or copilot fly or attempt to fly with fuel contaminated by water?	# TOTAL CONTAMINATED FUEL .... |_|_|_| **IF 0, SKIP TO ER12.**
	A. ASK ONLY IF MORE THAN ONE MAKE/MODEL IN A10. Which (airplane/helicopter) had water-contaminated fuel? Was it (READ A10 MAKE/MODEL LIST)?	SPECIFY MAKE/MODEL:

ER12. Fly or attempt to fly with the wrong type of fuel?

TOTAL WRONG FUEL .... |_|_|_|
IF 0, SKIP TO ER13.

A. ASK ONLY IF MORE THAN ONE MAKE/MODEL IN A10.

Which (airplane/helicopter) flew or attempted to fly with the wrong type of fuel? Was it (READ A10 MAKE/MODEL LIST)?

SPECIFY MAKE/MODEL:

ER13. Experience a failure of the attitude indicator or artificial horizon?

TOTAL ATTITUDE INDICATOR .... |_|_|_|
IF 0, SKIP TO ER14.

A. ASK ONLY IF MORE THAN ONE MAKE/MODEL IN A10.

Which (airplane/helicopter) experienced this failure? Was it (READ A10 MAKE/MODEL LIST)?

SPECIFY MAKE/MODEL:

B. (Of the [# in ER13] times the attitude indicator failed, how many occurred/Did this failure of the attitude indicator occur) in instrument meteorological conditions or I.M.C? I.M.C means the visibility was less than three miles and/or the ceiling was less than 1,000 feet above ground.

TOTAL ATTITUDE INDICATOR IN IMC........ |_|_|_|

INTRODUCTION:

My next questions relate to **turbulence**.

TU1. During the last 60 days, how many times did (an airplane/a helicopter) on which you were a pilot or copilot encounter severe turbulence that caused large abrupt changes in altitude, airspeed, or attitude

CAUSED ABRUPT CHANGES........ |_|_|_|
IF 0, SKIP TO TU2.

A. (Of the [#in TU1] severe turbulence encounters, how many occurred/Did this severe turbulence encounter occur) in I.M.C. conditions? [Note to interviewer: I.M.C. = INSTRUMENT METEOROLOGICAL CONDITIONS]

IN IMC CONDITIONS .... |_|_|_|

THE AMOUNT IN TU1A CANNOT BE GREATER THAN THE AMOUNT IN TU1.

B. (Of the [# in TU1] severe turbulence encounters, how many occurred/Did this severe turbulence encounter occur) in clear air?

IN CLEAR AIR........ |__|__|__|

THE AMOUNT IN TU1A AND TU1B CANNOT BE GREATER THAN THE AMOUNT IN TU1.

C. (Of the [# in TU1] severe turbulence encounters, how many resulted in one or more occupants being injured.

INJURY EVENTS........ |__|__|__|

TU2. Encounter wake turbulence that resulted in 45 or more degrees of aircraft roll

RESULTING IN AIRCRAFT ROLL........ |__|__|__|

INTRODUCTION:

My next questions are about **weather-related events while airborne**.

WE1. During the last 60 days, how many times did (an airplane/a helicopter) on which you were a pilot or copilot lack accurate weather information when you needed it while airborne?

LACK WEATHER INFORMATION........ |__|__|__|

IF 0, SKIP TO WE2.

A. (Of the [# WE1] times when you lacked accurate weather information, how many involved non-U.S. airports or controllers? Did this time when you lacked accurate weather information involve a non-U.S. airport or controller?)

INVOLVE NON-US AIRPORT OR CONTROLLER... |__|__|__|

THE AMOUNT IN WE1A CANNOT BE GREATER THAN THE AMOUNT IN WE1.

B. (Of the [# WE1] times when you lacked accurate weather information, how many involved ATIS? Did this time when you lacked accurate weather information involve ATIS?)

INVOLVE ATIS........ |__|__|__|

THE AMOUNT IN WE1B CANNOT BE GREATER THAN THE AMOUNT IN WE1.

C. (Of the [# WE1] times when you lacked accurate weather information; how many involved Flight Service Stations (FSS)? Did this time when you lacked accurate weather information involve a Flight Service Station (FSS?)

INVOLVE FSS........ |__|__|__|

THE AMOUNT IN WE1C CANNOT BE GREATER THAN THE AMOUNT IN WE1.

D. (Of the [# WE1] times when you lacked accurate weather information, how many involved Flight Watch? Did this time when you lacked accurate weather information involve Flight Watch?)

INVOLVE FLIGHT WATCH........ |__|__|__|

THE AMOUNT IN WE1D CANNOT BE GREATER THAN THE AMOUNT IN WE1.

E. (Of the [# WE1] times when you lacked accurate weather information, how many involved the Automatic Weather Observation Service (AWOS) or Automatic Surface Observation Service (ASOS)? Did this time when you lacked accurate weather information involve the Automatic Weather Observation Service (AWOS) or Automatic Surface Observation Service (ASOS)?)

INVOLVE AWOS .... |__|__|__|

THE AMOUNT IN WE1E CANNOT BE GREATER THAN THE AMOUNT IN WE1.

WE2-A. AIRPLANE ONLY.
How many times did an airplane divert to an alternate airfield because of weather?

DIVERT TO ALTERNATE AIRFIELD .... |__|__|__|
AIRPLANE GO TO WE3-A.

WE2-H HELICOPTER ONLY.
How many times did a helicopter divert to an alternate airfield, heliport or land because of weather?

DIVERT TO ALTERNATE AIRFIELD .... |__|__|__|
HELICOPTER GO TO WE3-H.

WE3-A AIRPLANE ONLY.
How many times did an airplane experience airframe icing that reduced the aircraft's ability to maintain altitude, speed, stability, or directional control?

EXPERIENCE AIRFRAME ICING .... |__|__|__|

WE3-H HELICOPTER ONLY.
How many times did a helicopter experience airframe or rotor icing that reduced the aircraft's ability to maintain altitude, speed, stability, or directional control?

EXPERIENCE AIRFRAME ICING .... |__|__|__|

WE4. During the last 60 days, how many times did (an airplane/a helicopter) on which you were a pilot or copilot encounter windshear or a microburst condition that resulted in an airspeed deviation of 15 knots or greater?

ENCOUNTER WINDSHEAR/MICROBURST .... |__|__|__|
AIRCRAFT SKIP TO CP1. HELICOPTER CONTINUE.

WE5-H HELICOPTER ONLY.
How many times did a helicopter experience loss of tail rotor effectiveness due to high density altitude?

ROTOR EFFECTIVENESS ALT .... |__|__|__|

WE6-H HELICOPTER ONLY.
How many times did a helicopter experience loss of tail rotor effectiveness due to high winds?

ROTOR EFFECTIVENESS WINDS .... |__|__|__|

WE7-H HELICOPTER ONLY.
How many times did a helicopter experience loss of the visible horizon due to white out or brown out conditions on either takeoff or landing?

INVOLVE NON-US AIRPORT OR CONTROLLER ... |__|__|__|

INTRODUCTION:

My next question is about **passenger-related events.**.

CP1. During the last 60 days, how many times were you distracted by a passenger while in flight through conversation or physical contact?

NOTE TO INTERVIEWERS: INCLUDES TAPPING ON SHOULDER.

PAX DISTRACT .... |__|__|__|

INTRODUCTION:

My next questions are about **airborne conflicts.** Just as a reminder, we are only asking about events that you experienced flying under FAR Part 135 or Part 91 as (an airplane/a helicopter) pilot or copilot.

How many times did (an airplane/a helicopter) (READ QUESTION)?

AC1. Experience a bird strike?

BIRD STRIKES .... |__|__|__|

AC2. Perform an evasive action to avoid an imminent in-flight collision with another aircraft that was never closer than 500 feet?

EVASIVE ACTIONS .... |__|__|__|

AC3. Experience less than 500 feet of separation from another aircraft while both aircraft were airborne?

LESS THAN 500 FEET SEPARATION .... |__|__|__|

INTRODUCTION:

The next few questions are about **ground operations**.

GE1. During the last 60 days, how many times did (an airplane/a helicopter) on which you were a pilot or copilot Land at a location without a wind sock, wind vane, or other wind indicator device?

WIND INDICATOR .... |__|__|__|

GE2.	Take off, or attempt to take off, with control locks, pitot covers, or other protective gear still attached to the aircraft? NOTE TO INTERVIEWER: INCLUDES BUT NOT LIMITED TO: GEAR FLAGS; ENGINE, INTAKE, OR EXHAUST PLUGS; TIE-DOWNS.	# PROTECTIVE GEAR........ |_|_|_|
GE3.	Experience an unplanned aborted or rejected takeoff?	# REJECTED TAKEOFFS........ |_|_|_| HELICOPTER SKIP TO GE8. AIRPLANE CONTINUE.
GE4-A.	AIRPLANE ONLY. During the last 60 days, how many times did (an airplane/a helicopter) on which you were a pilot or copilot go off the edge of a runway or taxiway while taxiing?	# GO OFF EDGE RUNWAY/TAXIWAY........ |_|_|_|
GE5-A.	AIRPLANE ONLY. Go off the **edge** of a runway while taking off or landing?	# GO OFF EDGE OF RUNWAY........ |_|_|_|
GE6-A.	AIRPLANE ONLY. Go off the **end** of the runway?	# GO OFF END OF RUNWAY........ |_|_|_|
GE7-A.	AIRPLANE ONLY During the last 60 days, how many times did (an airplane/a helicopter) on which you were a pilot or copilot inadvertently enter an active runway?	# ENTER ACTIVE RUNWAY .... |_|_|_|
GE8-A	AIRPLANE ONLY Begin takeoff while another aircraft occupied or was crossing the same runway?	# TAKEOFF ROLL WITH OCCUPIED RUNWAY........ |_|_|_|
GE9-A	AIRPLANE ONLY Land while another aircraft occupied or was crossing the same runway?	# LAND ON OCCUPIED RUNWAY........ |_|_|_| HELICOPTER SKIP TO GE11. AIRPLANE CONTINUE.
GE10-A.	AIRPLANE ONLY. Hit or collide with a runway or taxiway light?	# HIT LIGHTS........ |_|_|_|
GE11.	During the last 60 days, how many times did (an airplane/a helicopter) on which you were a pilot or copilot hit a deer or other animal other than a bird?	# HIT ANIMAL .... |_|_|_| HELICOPTER SKIP TO GE13. AIRPLANE CONTINUE.

GE12-A. AIRPLANE ONLY.
Collide or nearly collide with a ground vehicle?

COLLIDE WITH GROUND VEHICLE .... |__|__|__|
IF 0, SKIP TO GE14.

A. (Of the [# in GE12] near collisions with a ground vehicle, how many occurred/Did this near collision with a ground vehicle occur) while your aircraft was on the ramp or apron?

ON RAMP/APRON/GATE AREA .... |__|__|__|
THE AMOUNT IN GE12A CANNOT BE GREATER THAN THE AMOUNT IN GE12.

B. (Of the [# in GE12] near collisions with a ground vehicle, how many occurred/Did this near collision with a ground vehicle occur) while your aircraft was on the taxiway?

ON TAXIWAY .... |__|__|__|
THE AMOUNT IN GE12A AND GE12B COMBINED CANNOT BE GREATER THAN THE AMOUNT IN GE12.

C. (Of the [# in GE12] near collisions with a ground vehicle, how many occurred/Did this near collision with a ground vehicle occur) while your aircraft was on the runway?

ON RUNWAY .... |__|__|__|
THE AMOUNT IN GE12A, GE12B, AND GE12C COMBINED CANNOT BE GREATER THAN THE AMOUNT IN GE12.

SKIP TO GE14.

GE13-H. HELICOPTER ONLY.
Collide or nearly collide with a ground vehicle?

COLLIDE WITH GROUND VEHICLE .... |__|__|__|
.... IF 0, SKIP TO GE15.

A. (Of the [# in GE13] near collisions with a ground vehicle, how many occurred/Did this near collision with a ground vehicle occur) while your aircraft was operating at an airport, **not** a heliport?

AT AIRPORT .... |__|__|__|
THE AMOUNT IN GE13A CANNOT BE GREATER THAN THE AMOUNT IN GE13.

B. (Of the [# in GE13] near collisions with a ground vehicle, how many occurred/Did this near collision with a ground vehicle occur) while your aircraft was operating at a heliport? NOTE TO INTERVIEWER, NOT AT AN AIRPORT.

AT HELIPORT .... |__|__|__|
THE AMOUNT IN GE13A AND GE13B COMBINED CANNOT BE GREATER THAN THE AMOUNT IN GE13.

C. (Of the [# in GE13] near collisions with a ground vehicle, how many occurred/Did this near collision with a ground vehicle occur) while your aircraft was operating at an unprepared landing site?

UNPREPARED SITE .... |__|__|__|
THE AMOUNT IN GE13A, GE13B, AND GE13C COMBINED CANNOT BE GREATER THAN THE AMOUNT IN GE13.

SKIP TO GE15.

GE14-A. AIRPLANE ONLY.
During the last 60 days, how many times did (an airplane/a helicopter) on which you were a pilot or copilot nearly experience a ground collision with another aircraft while both aircraft were on the ground?

NEAR GROUND COLLISION .... |__|__|__|
.... IF 0, SKIP TO GE15.

A. (Of the [# in GE14] near collisions with another aircraft, how many occurred/Did this near collision with another aircraft occur) while your aircraft was on the ramp or apron?

ON RAMP/APRON/GATE AREA .... |__|__|__|
THE AMOUNT IN GE14A CANNOT BE GREATER THAN THE AMOUNT IN GE14.

B. (Of the [# in GE14] near collisions with another aircraft, how many occurred/Did this near collision with another aircraft occur) while your aircraft was on the taxiway?

ON TAXIWAY .... |__|__|__|
THE AMOUNT IN GE14A AND GE14B COMBINED CANNOT BE GREATER THAN THE AMOUNT IN GE14.

C. (Of the [# in GE14] near collisions with another aircraft, how many occurred/Did this near collision with another aircraft occur) while your aircraft was on the runway?

ON RUNWAY .... |__|__|__|
THE AMOUNT IN GE14A, GE14B, AND GE14C COMBINED CANNOT BE GREATER THAN THE AMOUNT IN GE14.

GE15. During the last 60 days, how many times did you experience a collision or near collision with anything other than an animal, a ground vehicle, or another aircraft while on the ground?

OTHER GROUND COLLISION .... |__|__|__|
.... IF 0, SKIP TO AH1.

A. What were the objects you collided with or nearly collided with?

SPECIFY: ____________________

INTRODUCTION:
My next questions are about **aircraft handling-related events.**

AH1. During the last 60 days, how many times did (an airplane/a helicopter) on which you were a pilot or co-pilot use some of its reserve fuel as defined by the FARs (Federal Aviation Regulations)?

USE RESERVE FUEL .... |__|__|__|

AH2. Accept an A.T.C. clearance that the (airplane/helicopter) could not comply with because of its performance limits?

ACCEPT CLEARANCE NOT COMPLY WITH .... |__|__|__|

AH3.	Lose sight of another aircraft from which the pilot or copilot was trying to maintain visual separation?	# LOSE SIGHT OF AIRCRAFT.................................... |_|_|_| **IF 0, SKIP TO AH4.**
	A. (Of the [# in AH3] times an aircraft lost sight of another aircraft, how many occurred/Did losing sight of another aircraft occur) in marginal visual conditions of 3 miles or less?	# IN MARGINAL VISUAL CONDITONS .... |_|_|_| ***THE AMOUNT IN AH3A CANNOT BE GREATER THAN THE AMOUNT IN AH3.***
AH4.	Inadvertently land without clearance at an airport with an active control tower	# LAND W/O CLEARANCE.. |_|_|_|
	As a reminder, these questions still refer to the last 60 days. During the last 60 days, how many times did an aircraft on which you were a pilot or co-pilot (READ QUESTION)?	
AH5.	During the last 60 days, how many times did (an airplane/a helicopter) on which you were a pilot or co-pilot inadvertently begin takeoff without A.T.C. (air traffic control) clearance at an airport with an active control tower?	# TAKEOFF ROLL W/O CLEARANCE .... |_|_|_|
AH6	Inadvertently deviate from an assigned routing or A.T.C. vector for one minute or more?	# DEVIATIONS... |_|_|_|
AH7.	Take off with an out-of-limit center of gravity?	# TAKE-OFF OUT-OF-LIMIT CENTER OF GRAVITY. |_|_|_|
AH8.	Take-off overweight?	# TAKE-OFF OVERWEIGHT .... |_|_|_| HELICOPTER SKIP TO AH10. AIRPLANE CONTINUE.
AH9-A.	AIRPLANE ONLY. Commence take-off roll with an improper aircraft configuration?	# WITH IMPROPER CONFIGURATION...................... |_|_|_|
AH10.	During the last 60 days, how many times did (an airplane/a helicopter) on which you were a pilot or co-pilot experience an unintended unusual attitude for any reason?	# UNUSUAL ATTITUDE.. |_|_|_| AIRPLANE SKIPTO AH11. HELICOPTER CONTINUE.
	As a reminder, these questions still refer to the last 60 days. During the last 60 days, how many times did an aircraft on which you were a pilot or co-pilot (READ QUESTION)?	
AH11-H	HELICOPTER ONLY. Experience a valid low rotor R.P.M warning for any reason?	# LOW RPM WARNING.. |_|_|_| HELICOPTER SKIP TO AH12. AIRPLANE CONTINUE.

AH11-A AIRPLANE ONLY.
Experience an unintentional stall or valid stall warning?

STALL WARNING/STICK SHAKER ACTIVATION ... |_|_|_|

AH12. Nearly collide with terrain or ground obstruction or wires while airborne?

NEAR COLLISIONS/GROUND .... |_|_|_|
IF 0, AIRPLANE SKIP TO AH13, HELICOPTER SKIP TO A14.

INTERVIEWER NOTE: INCLUDES BUILDINGS

A. (Of the [# in AH12] near collisions with terrain, ground obstruction or wires, how many were/Was this near collision with terrain, ground obstruction or wires)-brought to your attention by A.T.C.(Air Traffic Control)?

ATC BROUGHT TO YOUR ATTENTION .... |_|_|

THE AMOUNT IN AH12A CANNOT BE GREATER THAN THE AMOUNT IN AH12.

B. (Of the [# in AH12] near collisions with terrain, ground obstruction or wires, how many were/Was this near collision with terrain, ground obstruction or wires) detected through direct sighting of the ground or obstruction?

DETECTED THROUGH DIRECT SIGHTING .... |_|_|

THE AMOUNT IN AH12A AND AH12B COMBINED CANNOT BE GREATER THAN THE AMOUNT IN AH12.

C. (Of the [# in AH12] near collisions with terrain, ground obstruction or wires, how many involved just wires?

AH13-A AIRPLANE ONLY.
Inadvertently cross the runway threshold during the landing approach with the landing gear up?

CROSS WITH GEAR UP .... |_|_|

A. (Of the [# in AH13] times an aircraft crossed the runway threshold with the landing gear up, how many times/The time the aircraft crossed the runway threshold with the landing gear up,) did you land with the gear up?

LAND WITH GEAR UP.. |_|_|

AH14. During the last 60 days, how many times did (an airplane/a helicopter) on which you were a pilot or co-pilot inadvertently enter airspace the aircraft was not cleared for?

UNCLEARED AIRSPACE .... |_|_|

As a reminder, these questions still refer to the last 60 days. During the last 60 days, (READ QUESTION)?

AH15. How many times did you lose track of the natural horizon due to reduced visibility while flying under Visual Flight Rules (V.F.R)?

LOSE HORIZON .... |_|_|

INTRODUCTION:
The next few questions are about **altitude deviations**.

AD1. How many times during the last 60 days did (an airplane/a helicopter) on which you were a pilot or copilot **inadvertently** deviate from an altitude assigned by A.T.C by more than 300 feet?

ALTITUDE DEVIATIONS .... |__|__|__|

AD2. ASK ONLY IF A12 > 0. OTHERS SKIP TO AT1. Earlier, you indicated you flew A12 flights. How many times during these A12 IFR flights) did you descend below Minimum Safe Altitude when you were **not** following A.T.C. radar vectors?

NOT FOLLOWING ATC RADAR VECTORS .... |__|__|__|

INTRODUCTIONS:
The next few questions are about **interactions with air traffic control**.

AT1 During the last 60 days, how many times was (an airplane/helicopter) on which you were a pilot or copilot unable to communicate with A.T.C. in a time-critical situation because of frequency congestion?

UNABLE TO COMMUNICATE WITH ATC .... |__|__|__|
IF 0, SKIP TO AT2.

A. (Of these [# in AT1] times you were unable to communicate with A.T.C. in a time-critical situation because of frequency congestion, how many occurred/Did the time you were unable to communicate with A.T.C in a time critical situation because of frequency congestion occur) **while on the ground**?

WHILE ON GROUND .... |__|__|__|
TIMES

THE AMOUNT IN AT1A CANNOT BE GREATER THAN THE AMOUNT IN AT1.

B. (Of these [# in AT1] times you were unable to communicate with A.T.C. in a time-critical situation because of frequency congestion, how many occurred/Did the time you were unable to communicate with A.T.C in a time critical situation because of frequency congestion occur) **while airborne in the terminal area**?

WHILE AIRBORNE .... |__|__|__|
TIMES

THE COMBINED TOTALS IN AT1A AND AT1B CANNOT BE GREATER THAN 100.

C. (Of these [# in AT1] times you were unable to communicate with A.T.C. in a time-critical situation because of frequency congestion, how many occurred/Did the time you were unable to communicate with A.T.C in a time critical situation because of frequency congestion occur) **while en route**?

WHILE EN ROUTE .... |__|__|__|
TIMES

THE COMBINED TOTALS IN AT1A, AT1B, AND AT1C CANNOT BE GREATER THAN 100.

AT2. How many times did (an airplane/a helicopter) fly at an undesirably high altitude or airspeed on approach due to an A.T.C. clearance
(NOTE TO INTERVIEWERS: THIS INCLUDES BUT MAY NOT BE LIMITED TO "SLAM DUNK" APPROACHES.)

HIGH ALTITUDE OR AIRSPEED .... |__|__|__|

AT3.	How many times did (an airplane/a helicopter) leave a communications frequency with A.T.C to get a weather briefing?	# LEAVE FREQ FOR WEATHER .... |__|__|__|
AT4.	How many times during the last 60 days were you informed that (an airplane/a helicopter) on which you were a pilot or copilot missed a transmission from A.T.C?	# MISS TRANSMISSION .... |__|__|__|
	A. Of the [# in AT4] times you missed a transmission from ATC, how many occurred/Did the time you missed a transmission from A.T.C occur due to being on the wrong frequency?	# WRONG FREQUENCY .... |__|__|__|
	B. Of the [# in AT4] times you missed a transmission from ATC, how many occurred/Did the time you missed a transmission from A.T.C occur due to high cockpit noise?	# COCKPIT NOISE |__|__|__|
	1) Were you wearing a communication headset at the time? (Note to interviewers: This includes helmets with integral headset speakers)	# HEADSET .... |__|__|__|
AT5.	How many times did you receive out of date, inaccurate or no information about relevant NOTAMs? NOTE TO INTERVIEWERS: NOTAMS = "NOTICES TO AIRMEN"	# NOTAMS .... |__|__|__|

General Aviation Questionnaire

Section C: Weather-related Issues

SECTION C: WEATHER-RELATED ISSUES

INTRODUCTION:

The next set of questions is a special focused topic section of the survey regarding weather-related issues. The first set of questions asks about **weather planning for your flights.** Just as a reminder, we are still only asking about events that you experienced flying under FAR Part 135 or Part 91 as (an airplane/a helicopter) pilot or copilot. Again, we use the terms "flight" throughout this interview to mean the period of time between each takeoff and landing, even if that time is short such as for instructors teaching students to land.

THE FOLLOWING QUESTIONS ARE FOR ALL PILOTS

C1. Earlier in the interview, you indicated you made [# TAKEOFFS IN A5a+A6a for airplane or a5b and a6b for helicopter] takeoffs in (an airplane/a helicopter) during the past 60 days. (For how many of these flights did you obtain pre-flight weather information? /On this flight, did you obtain pre-flight weather information?)

FLIGHTS WEATHER BRIEFING .... |__|__|__|

IF 0, SKIP TO C4.

A. (Of these [# FLIGHTS C1] flights where you obtained preflight weather information, how many were obtained by/Was the preflight weather information obtained by) (READ QUESTIONS):

1. **Commercial TV, radio, or cable weather broadcast** that was **not specific to aviation**?

   # COMMERCIAL NON-AVIATION .... |__|__|__|
   CANNOT BE GREATER THAN C1.

2. **Commercial TV, radio, or cable weather broadcast** that was **specific to aviation?**

   # COMMERCIAL AVIATION .... |__|__|__|
   CANNOT BE GREATER THAN C1

3. **Company provided weather from a dispatcher**?

   # COMPANY............ |__|__|__|
   CANNOT BE GREATER THAN C1

4. **DUATS or other computer-accessed aviation weather service**?

   DUATS = computer-based weather service provided by the F.A.A.

   # COMPUTER ACCESS .... |__|__|__|
   CANNOT BE GREATER THAN C1

5. **Pre-recorded Flight Service Station Weather Brief**?

   Flight Service Station = F.S.S.

   # PRE-RECORDED .... |__|__|__|
   CANNOT BE GREATER THAN C1

6. **Verbal briefing with an FAA flight service station specialist** (F.S.S)?

   # VERBAL BRIEFING .... |__|__|__|
   CANNOT BE GREATER THAN C1

	7. Did you obtain pre-flight weather information in some other way?	YES ....1 NO .... (SKIP TO C2) ....0 RF .... (SKIP TO C2) ....7 DK .... (SKIP TO C2) 8
	a. How did you obtain the weather information?	SPECIFY:
C2.	IF ONLY ONE QUESTION ANSWERED IN C1A1-7, SKIP TO C2A. You said you used the following pre flight weather information sources in the last 60 days (LIST ITEMS CODED ONE OR HIGHER IN C1A1-7. Which did you use most recently?	COMMERCIAL NOT SPECIFIC TO AVAITION COMMERCIAL SPECIFIC TO AVIATION COMPANY PROVIDED DUATS OR OTHER COMPUTER ACCESSED PRE-RECORDED FLIGHT SERVICE STATION VERBAL FAA BREIFING OTHER RF DI(
	A. How understandable was the weather information you received most recently from (SOURCE LISTED IN C2)? Would you say it was [READ OPTIONS]?	Not at all understandable ....1 Slightly understandable ....2 Moderately understandable ....3 Very understandable ....4 Extremely understandable ....5 RF ....7 DK 8
	B. How accurate was that weather information you received most recently from (SOURCE LISTED IN C2) in relation to the weather conditions you encountered during flight? Would you say the information was [READ OPTIONS]?	Not at all accurate ....1 Slightly accurate ....2 Moderately accurate ....3 Very accurate ....4 Extremely accurate ....5 RF ....7 DK 8
	C. How much time elapsed between your most recent weather briefing and the time of takeoff?	HOURS ....|_|_|_| MINUTES... ....|_|_|_|
C3	In which state or states do you primarily fly? INTERVIEWER: RECORD UP TO 3 STATES USING 2-DIGIT STATE CODE LISTED BELOW. IF PILOT GIVES OTHER TYPE OF ANSWER (E.G., "NORTHEAST," RECORD.	STATE 1: STATE 2: STATE 3: OTHER (RECORD)

C4. As a reminder, we are still only asking about events that you experienced flying under FAR Part 135 or Part 91 as (an airplane/a helicopter) pilot or copilot (Of the [# in A5A+A6A FOR AIRPLANE OR A5B AND A6B FOR HELICOPTER] takeoffs you made during the last 60 days, how many of these flights were/Was the takeoff you made during the last 60 days) conducted under V.F.R flight rules?

TAKEOFFS UNDER VFR.. |_|_|_|

CANNOT BE GREATER THAN A5A+A6A FOR AIRPLANE OR A5B AND A6B FOR HELICOPTER

VFR = Visual Flight Rules: Visibility greater than 3 miles and ceiling greater than 1,000 feet above ground level

C5. Do you, or your organization, apply pre-flight **V.F.R** weather minimums that are more conservative than those required by the F.A.A?

YES 1
NO .... (SKIP TO C6).......................................0
RF............................. (SKIP TO C6).......................................7
DK............................. (SKIP TO C6).......................................8

INTERVIEWER: IF PILOT MENTIONS IFR HERE, LET HIM/HER KNOW WILL BE GETTING TO IFR LATER IN THE INTERVIEW.

A. Under those more conservative weather minimums, what is the minimum number of **miles of visibility** you or your organization require?

MILES VFR MIN VISIBILITY .... |_|_|_|

B. Under those more conservative weather minimums, what is the minimum **ceiling in feet** that you or your organization require?

FEET VFR MIN CEILING .... |_|_|_|

INTRODUCTION: My next questions are about the weather related issues **during** the flights.

C6. Earlier in the interview, you indicated you made [# TAKEOFFS IN A5A+A6A FOR AIRPLANE OR A5B AND A6B FOR HELICOPTER] takeoffs as (an airplane/a helicopter) pilot or copilot during the past 60 days. (On how many of these flights/On that flight) did poor weather result in you loosing track of your position?

LOST DUE TO WEATHER .... |_|_|_|
CANNOT BE GREATER THAN A5+A6
IF 0, SKIP TO C7.

A. (For that time/For the most recent time that happened), what was the visibility in miles?

VISIBILITY IN MILES .... |_|_|_|
NEED TO ADD DECIMAL

C7. (In how many of the [# TAKEOFFS IN A5A+A6A FOR AIRPLANE OR A5B AND A6B FOR HELICOPTER] flights did you experience spatial disorientation from poor visibility due to weather?/On the one flight you made during the last 60 days, did you experience spatial disorientation from poor visibility due to weather?)

TIMES SPATIAL DISORIENTATION .... |_|_|_|

CANNOT BE GREATER THAN A5A+A6A FOR AIRPLANE OR A5B AND A6B FOR HELICOPTER.
IF 0, SKIP TO C8.

	A. For (that/the most recent) event, what was the estimated visibility in miles?	VISIBILITY IN MILES .... |__|__|__| NEED TO ADD DECIMAL
	B. (How many occurred at night?/Did that flight occur at night?	# SPATIAL DISORIENTATION AT NIGHT................ |__|__|__| **CANNOT BE GREATER THAN C7.**
C8	During the last 60 days, (on how many of the [# TAKEOFFS IN A5A+A6A FOR AIRPLANE OR A5B AND A6B FOR HELICOPTER] flights) how many times did you inadvertently enter instrument meteorological conditions or I.M.C while on a VFR flight? IMC = Instrument meteorological conditions: Visibility less than 3 miles and/or cloud ceiling less than 1,000 feet above ground level VFR = Visual Flight Rules: Visibility greater than 3 miles and ceiling greater than 1,000 feet above ground level	# INADVERTENT IMC.. |__|__|__| **CANNOT BE GREATER THAN A5A+A6A FOR AIRPLANE OR A5B AND A6B FOR HELICOPTER.** **IF 0, SKIP TO C9.**
	A. (How many times did this/Did this) occur at night	# IMC AT NIGHT.. |__|__|__| **CANNOT BE GREATER THAN C8.**
	B. How did you resolve [that/the most recent] inadvertent I.M.C problem? Did you (READ ANSWERS)?	Ask for A.T.C help without declaring an emergency. Ask for A.T.C help and declare an emergency. Reverse course Climb Descend File IFR Do something else (RECORD)________________________ RF DK
C9	(On how many of the [# TAKEOFFS IN A5A+A6A FOR AIRPLANE OR A5B AND A6B FOR HELICOPTER] flights/On the one flight you conducted during the last 60 days) did weather conditions result in you conducting a go-around or missed approach on landing?	# GO AROUND .... |__|__|__| **CANNOT BE GREATER THAN A5A+A6A FOR AIRPLANE OR A5B AND A6B FOR HELICOPTER.** **IF 0, SKIP TO C10.**
	A. How many were/Was this) due to due to poor visibility?	# GO AROUND VIS.. |__|__|__| **CANNOT BE GREATER THAN C9.**
	B. How many were/Was this) due to high winds?	# GO AROUND WINDS .... |__|__|__| **CANNOT BE GREATER THAN C9.**
C10	(On how many of the [# TALKEOFFS IN A5A+A6A FOR AIRPLANE OR A5B AND A6B FOR HELICOPTER] flights/On the one flight) you conducted during the last 60 days, how many times did worsening weather conditions result in you diverting to an alternative landing site?	# LAND DUE TO WEATHER .... |__|__|__|

A. On the most recent/On that) flight how did you determine that the weather was worsening? Did you (READ ANSWERS)?	RECEIVE AN UPDATED INFLIGHT WEATHER BRIEFING FROM A FLIGHT SERVICE STATION (ALL METHODS). OBSERVE THE WEATHER DIRECTLY FROM COCKPIT. OBTAIN PILOT REPORTS FROM OTHER PILOTS USING FLIGHT WATCH DO SOMETHING ELSE (RECORD) ________

The following questions are for VFR rated pilots only

(Determined from question A1=no)

All other skip to C15.

C11	My next questions are about instrument flying you may have conducted. Now I'm going to ask a few questions about instrument flying you may have conducted as (an airplane/a helicopter) pilot or copilot over the last 60 days. (On how many of the [# TAKEOFFS IN A5A+A6A FOR AIRPLANE OR A5B AND A6B FOR HELICOPTER] flights /On the one flight you conducted) did you find yourself flying V.F.R over a cloud deck sometimes called "V.F.R on top," where you had to penetrate the cloud deck in order to land? VFR = Visual Flight Rules: Visibility greater than 3 miles and ceiling greater than 1,000 feet above ground level	# VFR ON TOP.......... |_|_|_| **CANNOT BE GREATER THAN A5A+A6A FOR AIRPLANE OR A5B AND A6B FOR HELICOPTER**
	A. On the most recent/On that) flight how did you get through the cloud deck to land? Did you (READ CATEGORIES)?	ASK FOR ATC HELP WITHOUT DECLARING AN EMERGENCY. ASK FOR ATC HELP AND DECLARED AN EMERGENCY. DESCENDED THROUGH THE CLOUDS WITHOUT CONTACTING ANYONE FILE IFR OR SOMETHING ELSE RECORD) ________
C12	How many hours of instrument training have you received since you began to fly?	# HOURS OF INSTRUMENT TRAINING .... |_|_|_|
C13	How many hours of training have you received in actual I.M.C conditions (visibility less than three miles and/or ceiling less than 1,000 feet above ground level) since you began to fly? IMC = Instrument meteorological conditions: Visibility less than 3 miles and/or cloud ceiling less than 1,000 feet above ground level	# HOURS OF ACTUAL INSTRUMENT TRAINING ..... |_|_|_|

C14 How long ago was your last instrument training session?

NOTE: THIS INCLUDES BIENNIAL FLIGHT REVIEWS

YEARS ..... |_|_|_|

MONTHS ..... |_|_|_|

DAYS ..... |_|_|_|

The following questions are for IFR rated pilots only

(Determined from question A1=yes)

Other skip to D1

INTRODUCTION: My next questions are about instrument flying you may have conducted. Now I'm going to ask a few questions about instrument flying you may have conducted as (an airplane/a helicopter) pilot or copilot over the last 60 days.

C15 (On how many of the [# TAKEOFFS IN A5A+A6A FOR AIRPLANE OR A5B AND A6B FOR HELICOPTER] flights/On the one flight) did you file an I.F.R flight plan?

IFR: = Instrument Flight Rules

IFR FLIGHT PLANS ..... |_|_|_|

CANNOT BE GREATER THAN A5A+A6A FOR AIRPLANE OR A5B AND A6B FOR HELICOPTER
IF 0, SKIP TO C16.

A. Of these [# FLIGHTS IN C15] flights, how many had I.M.C conditions at least part of the time?/Did this flight have I.M.C conditions at least part of the time?)

#IMC CONDITIONS ..... |_|_|_|

CANNOT BE GREATER THAN C15

IMC: = Instrument meteorological conditions: Visibility less than 3 miles and/or cloud ceiling less than 1,000 feet above ground level

C16 Do you, or your organization, apply pre-flight **I.F.R** weather minimums that are more **conservative** than that required by the F.A.A?

IFR= Instrument Flight Rules

YES		1
NO	(SKIP TO C17)	0
RF	(SKIP TO C17)	7
DK	(SKIP TO C17)	8

A. Under those more conservative I.F.R weather minimums, what is the minimum number of **miles of visibility** you or your organization require?

IFR MILES VISIBILITY ..... |_|_|_|

B. Under those conservative I.F.R weather minimums, what is the **minimum ceiling in feet y**ou require?

IFR IN FEET CEILING ..... |_|_|_|

NOTE—MAY NEED MORE BLOCKS

C17	IF C15 IS 0, SKIP TO C18. During the last flight you flew where you filed I.F.R, did the aircraft have (READ QUESTIONS)?	
	A. Weather radar or thunderstorm detection equipment?	YES .... 1 NO .... 0 RF .... 7 DK 8
	B. Autopilot, including wing levelers?	YES .... 1 NO .... 0 RF .... 7 DK 8
	AIRPLANES ONLY C. Anti-icing equipment that is approved for flight in icing conditions?	YES .... 1 NO .... 0 RF .... 7 DK 8
C18	(On how many of the [# FLIGHTS IN C15] flights/On the one flight) you conducted during the last 60 days, did you fly an instrument approach to land in I.M.C.? IMC = Instrument meteorological conditions: Visibility less than 3 miles and/or cloud ceiling less than 1,000 feet above ground level	# INSTRUMENT LANDING IMC .... |_|_|_| .... I **F 0, SKIP TO D1** **CANNOT BE GREATER THAN C15**
	A. During the (last) flight where you flew an instrument approach to landing in I.M.C conditions, what type approach was flown?	ILS VOR RNAV GPS LDA SDB NDB BACK COURSE ILS OTHER__________
	B. During the (last) flight where you flew an instrument approach to landing in I.M.C conditions, what was the ceiling, in feet, during the approach?	# CEILING INSTRUMENT LANDING........ |_|_|_| DK RF
	C. During the last flight where you flew an instrument approach to landing in instrument meteorological conditions, what was the visibility during the approach in miles or RVR (NOTE: runway visual range)	# VISIBILITY INSTRUMENT MILES .... |_|_|_| RVR (RUNWAY VISUAL RANGE) IN FEET .... |_|_|_| DK RF

C19 Just as a reminder, we are only asking about events that you experienced flying under FAR Part 135 or Part 91 as (an airplane/ a helicopter) pilot or copilot.

You indicated that you made [# FLIGHTS C18] flights on which you conducted an instrument approach to landing in IMC during the last 60 days. (How many of these approaches were/Was this approach) conducted under FAR part 91?

INSTRUMENT PART 91 .... |__|__|__|

IF 0, SKIP TO D1

C20 As you may know, the F.A.A currently allows pilots flying under FAR Part 91 to conduct instrument approaches, but not landings, when the weather conditions at the instrument approach landing facility is below landing minimums.

A. Are you aware of these regulations?

YES/NO/RF/DK

B. You just indicated that you made [# FLIGHTS C19] instrument approach[es] in I.M.C and under FAR part 91 during the last 60 days. (How many of those times did you fly the/Did you fly that) approach with the reported weather conditions **below** the minimums for that approach as allowed by the F.A.A?

INSTRUMENT BELOW MIN .... |__|__|__|

DK

RF

C. (On the most recent/On that) approach did the airport have on-site weather reporting?

YES 1
NO 0
RF 7

DK..

D. (During how many of those approaches/ During the approach) was the weather above minimums when you landed?

INSTRUMENT BELOW MIN LAND.......................... |__|__|__|

General Aviation Questionnaire

Section D: Questionnaire Feedback

SECTION D: QUESTIONNAIRE FEEDBACK

INTRODUCTION:
I only have a couple more questions and these are about your reactions to the survey we have just done.

D1. How confident are you that you accurately counted all of the safety-related events that I asked you about? Would you say you were (READ QUESTIONS)?

- Not confident at all 1
- Slightly confident 2
- Moderately confident 3
- Very confident 4
- Extremely confident 5
- RF 7
- DK 8

D2. Were any of the questions I asked confusing, poorly worded, or ambiguous?

- YES 1
- NO .... (SKIP TO E3) 0
- RF .... (SKIP TO E3) 7
- DK .... (SKIP TO E3) 8

A. Could you please describe these question problems? RECORD VERBATIM. AT COMPLETION OF INTERVIEW, ENTER QUESTION NUMBER.

QUESTION NUMBER	RECORD VERBATIM

D3. Are there any safety problems happening within the national aviation system that I did not ask about but that you think may be worth asking about in further surveys?

- YES 1
- NO .... (SKIP TO E4) 0
- RF .... (SKIP TO E4) 7
- DK` .... (SKIP TO E4) 8

A. What are these problems?

SPECIFY: ______________________________

D4. Do **you** use the internet at home?

- YES 1
- NO 0
- RF 7
- DK 8

D5. Do you have any other comments or suggestions about this survey? RECORD VERBATIM.

ENDINT Again, thank you very much for your time and your help with this survey. Your input will help the aviation industry a great deal to measure the level of safety in the aviation system and will be held in confidence.

QUESTIONNAIRE LENGTH: QUESTIONNAIRE LENGTH (MINUTES) |__|__|__|